# CORNWALL'S MINES AND MINERS

# CORNWALL'S MINES AND MINERS

*nineteenth century studies by George Henwood*
*edited by Roger Burt*

# BRADFORD BARTON

TRURO

*3 1 2 4 6 4*

*HD*
*8039*
*M72 G 728*
*1972*

# CONTENTS

# INTRODUCTION

THESE PEN PORTRAITS OF ASPECTS OF DAILY LIFE AND LABOUR in Cornish mines and mining communities were drawn by George Henwood and were first published in the *Mining Journal* between June 1857 and May 1859 under the title 'Cornish Mine Photographs' and 'Cornish Mining Maxims'.[1] During these years Henwood was employed as Cornish correspondent of the *Journal* with the principal task of returning detailed intelligence on the condition of the major mines in the county as a basis for the daily transactions of the London and provincial mining markets. These cameos were provided as an extra light relief for the readers of an otherwise entirely formal and technical market weekly. By present standards they are often extravagantly verbose and overly effusive in style but they achieved great popularity and a wide following during their day. The quality of their content and contribution, however, is unquestionable. Taken together they represent one of the most perceptive and colourful commentaries on life in the mining communities ever to be compiled. Henwood turned the same astute and unblinking eye which he gave to his mine reports on the character and quality of life around him; recording its pleasures, criticising its injustices, and sketching its customs and past-times. He was essentially a man of his time and part of the scene which he surveyed, interpreting his world in terms of a rigid code of mid-Victorian middle-class ethics and morality tempered with a degree of humanity and liberalism. His portraits are as useful for the insight which they give into the minds of men of the period as they are for their descriptive detail. They lack the economic, technical, and geological analysis of William Pryce, Henry de la Beche, and Robert Hunt but tell in a sensitive and sympathetic way what it was like to be alive and involved then.

The clarity and insight of Henwood's writing was bred of a wide and diverse career which had brought him into first-hand contact with every strata of Cornish society. He was born into relatively humble circumstances at Penryn on the 25th September 1809, the son of Nicholas and Mary Henwood, landlords of the Western Hotel at Penzance. His early years were spent near Falmouth and he began his working life there as an apprenticed carver and gilder. However, like so many Cornishmen of his day, he was soon captivated by the adventure of mining and showing an ability and perseverance well beyond the average trained himself in the wide variety of theoretical and practical skills of mine engineering. He later extended his learning to include classics, philosophy, science, and theology, developing a breadth of knowledge clearly reflected in these photographs and in several other articles and monographs on aspects of geology and tin streaming. His interests, however, were not entirely academic. As a working mine captain he invented and improved new types of dressing machinery for copper, lead, and zinc blende and he often advised others on day-to-day mining operations. Together perhaps with the influence and assistance of his cousin William Jory Henwood—a Fellow of both the Royal and Geological societies and one of the most eminent mining authorities of his day—this proven ability earned George Henwood the management of the Marquis of Breadalbane's extensive mining property in Scotland. This was undoubtedly a happy and satisfying chapter in his life and he was to write with admiration of the Marquis as one of the most adventurous mining promoters of his day. With his Scottish successes behind him he returned to Cornwall as correspondent of the *Mining Journal*. At this time he appears to have lived in the vicinity of Wendron village which provides the backdrop for many of his sketches. In the late eighteen sixties, at the age of nearly 60, he again left his native Cornwall, this time to manage mines in India worked by an English company in connection with the Maharajah of Puttialah. On his return to England some years later he settled at Darlaston in Staffordshire where he died aged 73 on the 27th March 1883. His latter years were saddened by the death of his second son, Cornwall Henwood, who had followed his father's career by working his way up from miner, to agent, and finally to mine promoter. Cornwall died in New York in 1881 having recently floated a company to work phosphate deposits in the Caribbean.

George Henwood's life and career were neither unique nor particularly remarkable. Indeed he was almost distinguished by his lack of distinction. However, it is perhaps this very ordinariness that best justifies his role as an observer of the world in which he lived.

*Roger Burt*
University of Exeter

---

1They were followed by a further series which is not included in this collection entitled "Photographs from Manufacturing Districts".

## SETTING AND PAY-DAY

"WHERE ARE ALL THOSE SMARTLY-DRESSED MEN, AND TIDY-looking lasses wending their way?" enquired two pedestrian tourists, as they, in the course of their journey through Cornwall, passed through Sticker Bridge, on their route to St. Austell. "Is there a fair or a wake, or anything of that kind, to be seen in the neighbourhood?" "Lor bless ye, no, Sirs" was the reply, "'tis settin and pay-day down to mine." "Pray what do you mean by setting and pay?" "Mean? why gets our wages for last month, and the men takes fresh bargains, to be sure!" "Wasn't you never to mine 'pon pay-day?—the Kappen wud make ye welcome if you are strangers."

Being desirous of learning everything connected with mining (as was their errand), after a little hesitation the pedestrians resolved they would go down—which they found to be up—to witness the ceremonies and customs enacted on setting and pay-day, at the Great Hewas Mines. If the reader please, he may accompany them.

Well, then, arriving on the mine before the setting commences, they amuse themselves by going into the great engine-house, and examining a remarkably fine, highly-finished, 60-inch engine, built by the late John Hodge, of St. Austell (a really splendid engine it is). Then the steam-whim, the stamps, and admirably laid out dressing-floors, attract their attention. The great number of well-dressed, civil, stalwart youths, and middle-aged men, with a moderate sprinkling of old ones, are seen quickening their pace towards the door of the humble counting-house, intimating the appointed time for business. One o'clock is approaching.

Now, all being assembled, the worthy captain appears, attended by the secretary, with book in hand, and waited on by a man holding a small box, containing a number of mysterious-looking small pebbles. The captain mounts an elevated situation, and after some brief remarks, proceeds to describe a pitch, or piece of work, to be contracted for, soliciting a price from the miners for its execution; when, after sundry questions, answers, and preliminary bids, up goes one of the mysterious pebbles into the air from the hand of the captain; any better offers must be made ere the stone falls to the ground (by the way a kind of Dutch auction). The lowest offer, if it suit the agent's views, is accepted. The contractor afterwards signs an agreement for himself and co-partners, and the bargain is concluded. If any of the men refuse to contract, the agents name their price, which usually elicits, "Well, Kappen, put en down, but 'tis too little for men to live." This ceremony is repeated for every tribute and tutwork pitch in the mine. The utmost decorum and respect for the managers and towards each other is observed. The men, as they take their bargains, retire, to calculate their chances for the ensuing month; the taker often to be blamed for having taken it too low, which he generally gets over by asserting that some one else would have taken the pitch, and "half a loaf is better than no bread."

The settings being ended, we see the modest, blushing, neatly clad, "bal maidens," as they are called, collecting around the pay window; the little "stamps," boys and girls, hop, skip, jump, and play all sorts of gambols on this, to them, monthly day of pleasure, when they are sure of a few halfpence, and "some meat for dinner to-morrow." The captains, purser, doctor, and upper servants having helped themselves to their salaries, and the cheques for the merchants' bills having been drawn, the "pares" of men (the term applied to a set of six or eight), who work together as partners, are called up. The man who took the setting for the last month appears, and having seen the account is correct, takes the amount and divides it amongst his fellows. The secretary, a worthy fellow, looks at the man through his spectacles, asks his name, though he knows it as well as the captain's, and then writes down, under the column "to whom paid," Nicholas Andrew, for self and partners.

After all the tributers and tutworkers are paid, the day men, labourers, and "maidens" get their wages; at last come our jolly little friends, some not more than 10 years of age, and some even younger than that; and it really does one's heart good to see the glee and delight, the consciousness of taking home "silver to mother" gives these youngsters. The names of all recipients are duly recorded. The order and method adopted render the whole affair the work of a few minutes.

The miners being dispatched, leave the mine (pay-day being always a holiday). One of the most important businesses of the day commences; the captain comes out and invites the strangers to walk in (if not too many), enquires names, and requests the favour of their company to dine at the Great Hewas Mine, an invitation our visitors were not loth to accept. The introduction of the purser, doctor, the land proprietor of a part of the Hewas Mines (Wheal Elizabeth), who resides close by, and looks well after their interests, to a few other large adventurers, and to the captains of two or three adjoining mines, take place. A neat, tidy, elderly woman appears to lay the cloth, which is soon done in the plainest possible manner. Savory smells assail the nose, provoking an already keen appetite, seeing it is three o'clock; and the neat little woman re-appears with a goodly, plain joint of roast beef, accompanied by potatoes and cabbage. The captain officiates, after grace is said (never omitted), and carves as if he were cutting for heroes. A jar of beer, vile stuff withal, from the public house two miles off, affords the bibatory at dinner, a little bread and cheese completes the welcome, grateful, and humble repast. Thanks having been again offered, and the cloth drawn, the captain produces two black bottles, which he soon informs you contain brandy and gin. "Take which you please, gentlemen, we must drink success to Great Hewas," which being, of course, duly honoured, the captains of the neighbouring mines, with the agents and underground captains of Hewas, hold a long discussion, and seek each other's counsel and advice as to the state and prospects of the mine generally, the conditions of the ends and levels, the quantity and price of tin, what improvements they have respectively made in the dressing processes, where they get the best and cheapest timber, coals, candles, &c., who supplies the best shovels and pickhilts, where they can procure good, able, steady, miners, &c., eliciting an amount of information nothing but a good dinner and good temper (everybody is in a good temper on pay-day) can command. On his health being drunk, the doctor rises, cracks a joke or two, says he is either glad he pockets his fees for doing nothing, as nobody is hurt, or that it is well two arms or legs were not broken instead of one, and that the accident is no worse. The adventurers, in turn, toast the captain, and praise him for his persevering endeavours, the secretary for his attention, and old Mr. Smith admonishes the captain to push on the works at Wheal Elizabeth. The captain proposes, "Welcome strangers," when the pedestrians rise together to thank the captains, secretary, adventurers, and all present, for their hospitality and society, and are about to leave, when the doctor politely enquires, "Where is your conveyance?" On being assured Johnny Foot's horse is the conveyance, he exclaims, "Nonsense! You cannot, and shall not, walk to St. Austell this evening. One can ride with me, and one with the purser," which politeness is again accepted.

After a most agreeable ride, and still more agreeable companions, our tourists arrive at the hostelry of Mr. Dunn, heartily delighted with all they have heard, seen, and experienced at the setting and pay-day of the Great Hewas Mine.

## THE MINER'S FUNERAL

"HARK! WHAT MEANS THAT DISTANT DEEP MELODY, WAFTING o'er this wild common?" exclaimed our tourists, one sultry Sunday afternoon, as they pursued their walk over the down near Stithian's "church-town;" and "what means that black mass descending yonder hill side?" enquired they of some stalwart, neatly-dressed youths, who were wending their way at their utmost speed to join the approaching throng. "That! why, that is a berrin—a miner's berrin.[1] Poor Nicholas Hill's—poor fella!—the best man in the mine: everybody liked en. He was killed last Thursday by a scale[2] falling, and was lashed[3] to atoms. We were forced to bring en up in a blankit. His own wife didn't know en: poor thing, 'tis a whish't[4] job for her. We do all pity her: only married one year, and got one child. She've never spoke, nor bit, nor supped since. They say she'll break her heart; and if she do, his poor old mother will die too. 'Tis dreadful for her too; for her poor old man was killed up to the mine also, and it brings it all home fresh to her, like, poor old soul!"

On being requested, as the funeral was drawing near, these evidently sympathising youths slackened their speed, and continued their melancholy story, from which it appeared that Nicholas Hill was one of those good-natured spirits, active, intelligent, frank kind of fellows, so frequently to be met with among our mining population, who win the respect of their superiors and equals alike, neither he nor they knowing how or why, as he makes no effort to do so. But the miner is interrupted, and he can tell his story much more graphically and pathetically than our pen can indite.

"'Tis a bitter job for us all, sure enuff, I assure ye, gentlemen: all the three parishes is up about it. The mine owners have offered to pay for the berrin; Mister Michael Williams have sent her a sovereign; and Kappen Billy Martin (that's our mine kappen) have sent her two. Kappen Billy is always good to the sick and to widows. Nick, poor fellow, was underground in the 210 fathom level, when a scale of ground came away from the roof of the level, and crushed en as flat as a pancake. His comrade[5] was wheeling the trade (copper ore) out to shaft; and when he comed back, there was poor Nick— poor fellow!" (wiping his eyes)—"as dead as a herring. When Tom Trezise (that's his comrade) comed up to surface, he cuddent speak for five minutes, and then says he:—'Kappen Billy, Kappen Billy, how shall I tell Sarah (that's his wife, sir)?—he's dead! he's killed!' 'Where—how—when?' shouts Kappen Billy. Kappen Billy is a good feeling man, gentlemen: without waiting to go to counting-house, down he goes to the 210 like a shot, and takes men with en; but 'twas no use: the doctor was sent for, but 'twas all over. So the doctor goes to the preacher, and they both goes to poor Sarah. When they reached her door, she says, 'My God, my God, I know all you are come for—my Nick is killed!" took up her child, kissed it, and then fainted away. She've never spoke nor slept since. His poor dear old mother (Aunt Jane, we call her) bore it a deal better. The preacher and doctor did and said all they cud to comfort her; but, poor thing, there's no more comfort for her in this world."

By this time they had closed in with the procession, which they found consisting of several thousands of neatly, plainly clad persons of both sexes, the majority in respectable mourning, headed by the aforesaid methodist minister, who gave out a hymn by verses, which was admirably sung by the choirs of the different chapels in the neighbourhood. They, and they only, who have heard can appreciate the melody produced by so large a number of these skilful singers, and these formed the strains that had at first attracted their attention. After the singers (upwards of 80 male and female voices) came the good Kappen Billy, and the other agents of the mine (Tresavean), arranged in deep mourning. On either side walked the relays of bearers.[6] The distance was three miles to the church. After the body came, or rather was carried, the unconscious form of the bereaved widow, who gazed listlessly on all around her. Reason had evidently lost its seat for the time; a deep settled melancholy not to be described—a sorrow and grief unutterable—had usurped its place. Well, indeed, might these unsophisticated sons of toil beautifully say, "There is no comfort for her in this world!" The two male companions who supported her were also almost overcome by the scene; when they beheld the old broken-down grandmother carrying the unconscious offspring, and bedewing its innocent smiling cheeks with her heart-wrung tears, their fortitude evidently gave way, and they wept too. A long string of relations walked in melancholy array, the rear being made up of thousands of miners and rustics, employed in the extensive works adjacent.

Our tourists involuntarily joined the procession, as it continued its melancholy parade and harmony, until it reached the churchyard. The knell, which had been tolling for some time before, now ceased; the preachers and singers filed off; the clergyman commenced the exquisitely beautiful ritual of the Church, when in an instant every male in that vast assembly was bareheaded, and all was solemnity itself, until it came to the thrilling sentence, "Earth to earth, dust to dust." The first rattle of the sod on the coffin startled the bereaved widow from her stupor, and, uttering a shriek that will never be forgotten, she went into a most distressing hysterical fit, in which state she was removed to the parsonage.

The ceremony being concluded, the good captain, preacher, agents and a few friends, took their way home, not merely to talk of the merits and demerits of the deceased, or to recount the scenes of the day, but to try how to put the poor widow into some way of earning bread for herself and her child. The choirs repaired to the various public houses, and sung their hymns over again; the young people forgot their griefs, and solaced themselves with a stroll in the company of their companions; the children laughed, sported and jumped from hillocked grave to grave (in itself a moral lesson on the instability of life); the friends and sexton left the churchyard; and the funeral of poor Nicholas Hill, the Tresavean Miner, was over.

This is a true and not overwrought picture of these western barbarians, as they are sometimes reckoned. The characters are daguerreotyped from nature: not one is fictitious. We would ask, can the highest ranks, the most educated and polished circles of society—or do they—display finer feelings, more intense suffering, more sympathy, or genuine gratitude, than is attempted feebly to be depicted in this little description of the Cornish miner's character in its humbler phases? They are a fine set of worthy fellows, if well treated: they are kind to their fellows, generous to the distressed, and frequently display the highest qualities with which humanity is endowed.

## ENGINE OPENING DAY

"ARE YOU AWARE THAT YOU ARE INVITED TO A CEREMONIAL dinner to-morrow?" enquired one of our tourists, as he handed an

open letter to his friend on making his appearance at a late breakfast
at mine host's of the White Hart, after a hard day's scramble and
ramble among the mines and cliffs in the neighbourhood. "Indeed
I was not," replied he, after reading it: "the invitation is very brief,
truly, but it is given with so much genuine Cornish cordiality and
welcome, and on so interesting an occasion, that refusal is out of the
question." As we were favoured with a sight of the letter, we found
it was an invitation to dine at the Great Wheal C—— Consols
Mines, where on that day they were about to set to work for the
first time a very large and splendid engine, by one of the most
celebrated engineers in the county. It was intimated that their company would be expected, and no excuses taken; that some of the
principal proprietors would be at the White Hart by early morn to
meet Capt. D., the manager, Mr. G., the purser, and other officials
of the mine, where they would all partake of a *dejeuner a la fourchette*,
by way of a beginning, and immediately proceed with the engineer
and contractor of the mine, which lay some seven or eight miles
from St. A——, near the rustic village of N——.

Whilst penning an acceptance of the polite invite, the door of the
apartment was opened by Mr. G., the purser, who had been before
introduced to them. He frankly and unceremoniously began by
giving details of their projected proceedings, and without giving
time for apology, question, or reply, went on saying, "Now, be
sure and be ready early. We expect Sir John D. will take the chair,
the lord of the manor, the parson, two or three magistrates, and
several others will attend. But I must be sharp, and secure conveyances." After musing for a minute he concluded by saying—"By
Jove! we must have four carriages, at least; one of you can ride with
me, and one with the doctor, as before." Ere we could well understand what he alluded to, or consider our situations, he was off to
make his arrangements, and seemed as much concerned as if he were
preparing for a lord mayor's banquet. His greatest anxiety, however,
appeared to be whether the reporter of the *Mining Journal* would be
present. "It will be well worth their while if he is," said he, proudly;
"as we shall all of us order a few copies for ourselves, and to send
to our friends; the landlord says he will give him bed, board, and
conveyance." We were emphatically exhorted not to do too much
at it to-day, as tomorrow would be a "stunner."

At the White Hart all hands were alert by four o'clock in the
morning. The post and stable boys joked merrily, and groomed their
steeds with glee, the prospect of the coming stir enlivening even

them. The ostler touched his forelock, as he took the horses of the captain and others who had arrived to welcome the gentlemen from "up the country." The old waiter bestirred himself, as well as he was able; the pretty waitresses, and spruce chambermaids bustled and hurried to and fro in their smartest attire; there was "such a getting up stairs," as evidently bespoke a great occasion was about to take place. Mine host and hostess were superintending the "laying out" the breakfast in the long ball-room, and had only just completed their duties, when the sound of a horn heralded the approach of the "gents," who drove up to the door of the hotel in noble style. After due congratulations on the part of the captains and persons present, our tourists were formally introduced. Sundry baskets, from which protruded the legs of animals and tails of birds, and cases marked with foreign brands and cyphers, were handed down, from, and out of, the vehicles, and Mr. Boots, with a knowing leer of the eye, enquired were they to go up stairs too? on being told they were to go to the mine, the smile on his face said, as plainly as if he spoke, "I wish I was also."

Ample justice having been done to the really handsomely served repast (the captain had advised all present to take in a good tamping to begin with), the party entered fresh carriages, not forgetting the mysterious-looking boxes and baskets (the contents of which Master Boots vowed 100 to 1 he could name). The carriages forming the van, the rear was brought up by the strangers and their companions; the reporter, who had arrived, and the host, being the last, in an appropriate dog-cart. A merry party were they that day when they started off at a smart pace for the village of N——. Cigars, of course, having been lighted, on they went through a cloud of smoke and dust; many a joke and gibe were bandied from friend to friend, many a kind hope expressed that they in the rear might get to their journey's end before "half play;" and many a kind offer to assist if they would throw a rope to them. And so they joked and journied until the merry church bells and sounds of music announced their approach to N——, at the entrance of which a "band" of five musicians headed the procession to the Cornish Arms, playing as lustily as they could blow and strike. "See the conquering hero comes," the tail of the procession being made up of all the idlers, men, women, and children, the neighbourhood could furnish. Here again, as a matter of course, it was necessary to "wet the whistle," "wash down the dust," "cool the coppers," &c., and sherry, bitter beer, bottled stout, and other such refreshers, were in requisition.

The cavalcade now slowly approached the scene of action, the mine,—we say slowly, for the very sufficient reason the roads were in such a state of repair mine host expressed considerable anxiety for the safety of his steeds and springs, the gents for their necks, and the captain for the basket of champagne he was determined to "stand" himself on this great event. The musicians were delighted, as it gave them ample opportunity to display their talents, which, if not equal to Jullien's celebrated performances, were quite as meritorious, every one doing his utmost to please.

At length, after many "hair-breadth 'scapes," they all arrived safely at the mine, where they were received by the lord of the manor on which the mine is situated, the clergyman of the parish, the magistrates, and other invited guests, with three loud hurrahs, aided by the powerful lungs of some scores of miners, who appeared to come out of all sorts of places, in every variety of costume, as if by magic. The gents having been duly shown into the new large counting-house, which had been decked with laurel for the occasion, and having been introduced to the local celebrities, were compelled, *nolens volens*, by the managers to toast the Great Wheal C—— Consols in a bumper of champagne. The mine at surface and machinery having been examined and explained by the captain and engineer, they go to view the noble boilers, and to ascertain if the steam was up (it certainly was with some of the party, in more senses than one). The engine-man, as a matter of course, nearly frightened one half the company, by turning the trial cock on at full power to prove the steam there, at all events, was up. The engineer stepped forward, begged to know what the name of the engine was to be, and who the godfather. The lord of the manor requested it should be named after the Chairman of the company, and he would himself become the sponsor. "Then please gentlemen, to stand here, and when she makes the first stroke, throw this bottle of port wine at her, christen her, and wish her success." The engineer then disappeared, but in a few seconds a hissing sound and gurgling noise was heard, and a few exclamations of—"Now, then, my brave boys!" and the ponderous beam, more than 25 tons, moved up and down, the christening ceremony was performed, the company and miners cheered lustily, the "band" rolled their noisiest sounds, and, before the day was over, the triumph of man's ingenuity moved "like a thing of life." The now proud captain, and still more proud engineer, were complimented by the Chairman, and retired to swagger a little at the day's success. The Chairman and strangers accepted an invitation to visit the mansion house, and be introduced to the ladies, and view the grounds, gardens and pictures, until the mine bell announced "in a quarter of an hour dinner on the table," when all returned, and found a substantial repast of old English fare, with bread and vegetables provided by the mine adventurers; to which was added a second course, admirably cooked at the mansion-house. This course, together with the wines, produced in ample quantity, had been the contents of the mysterious boxes and hampers before mentioned. A splendid dessert, from the gardens of the mansion, completed a dinner that would have cast no disgrace on the most accomplished *restaurateur*. The ponderous joints being removed to impromptu tables laid in the carpenter's shop and saw house, soon disappeared before the efforts of a combined attack from the miners, drivers, and band, leaving scarcely a wreck behind. Poor fellows! they appeared to, and no doubt did, enjoy their treat, equally as well as their superiors. They amused themselves over their beer and cider (provided *ad libitum* by the proprietors), if not quite so rationally as the gentlemen, they did quite as noisily: much to their honour, not a single instance of inebriety or misconduct had to be complained of.

In the court-house the Chairman of the company, as a matter of course, was lord of the feast; on his right sat the vicar, on the left the captain, the "vice" devolved on the lord of the manor, supported by the engineer and purser of the mine. After the usual loyal and patriotic toasts were disposed of, the Chairman said he had now come to the business of the day; he expressed his delight at their proceedings, the whole of which he approved, dilated extensively on the abilities of the engineer, as that day exemplified; on the captain's integrity; on his own confidence in the mine; on the pleasure he felt at the presence of the worthy vicar, and so many of the surrounding gentry; but more particularly of that of their excellent vice, whose liberality all had that day witnessed and partaken of, and whose good health, long life, and prosperity, he earnestly desired, and concluded by calling on him for a song.

The worthy squire, desirous of being all things to all men, good humouredly complied, and in a brief speech assured them all he had and would do his duty to the mine as long as the company did their's. (Cheers). If the mine would not pay he would not require dues. (Cheers.) No, not he! he would not distress any company, or rob any individual; he had not as yet become a shareholder, but as he was fully convinced they were a company who would work the mine in a miner-like manner he could not refrain from joining the adventure, therefore he desired Mr. G., the purser, to enter his name in the cost-book for 100 shares. (Repeated cheers.) He sat down after proposing long life to the noble Chairman of the Great Wheal C——Consols Mines, and may he meet them at no distant period to declare dividends; he should return the singing compliment, as he felt assured so smiling a countenance as the noble Chairman possessed would most certainly follow example. The noble chair replied satisfactorily and briefly.

Everybody having toasted everybody, and everybody having sung, everybody seemed satisfied; the reporter was warned to give a full account of the day's proceedings in the *Journal*, hence this lengthened notice. The procession returned in the same order it came, save a little more uproarious glee, to the village, when the same motley assemblage were in attendance to welcome them; instead of bitter beer, "Hot within the cold without" were substituted, the party arrived at St. A—— in time to find the hostess red-hot with rage, to think her beautiful supper had been spoiled by waiting. No matter; this was dispatched, green tea and good coffee being in demand; the party retired at an "early hour,"(?) to dream of an engine opening day.

N.B.—Report says there was a "St. Aubyn's day," but as it is the reporter's province to chronicle nothing but facts that come under his immediate attention, no notice can be taken of it—indeed, had we witnessed it a veil would have been thrown over it, "circumstances considered."

## SAMPLING DAY

OUR TRAVELLERS, WITH WHOM WE STILL KEEP COMPANY, having received and accepted an invitation to be present on sampling day at a celebrated mine near Redruth—being determined to witness every ceremony and process they had an opportunity of doing—sallied forth from P——n at an early hour, arriving at Redruth about 10 a.m. They at once walked to the residence of Capt. R., the manager, who immediately laid down the programme of the day's proceedings. "First," said he, "we will go to Wheal B——t, see them sample, thence to Wheal B——r, and then to T——n, where you shall see the man-engine at work, and go underground 200 or 300 fms., if you like; then come back to B——t to dinner, and then to my house to tea; and that's our day's work, so we've no time to lose."

Capital! thought our tourists, who had scarcely dreamt of such an excellent opportunity of seeing some of the largest and richest mines in the county, under such auspices. As the worthy captain had won the confidence and respect (as he always does) of his acquaintances, the elder said, "Now, Capt. R., I shall consider myself as a Cornish miner to-day, and comport myself as such." "Agreed," replied the captain, "now don't forget". Arrived at the mine—a walk of a mile or two—an introduction to Capt. P., the underground agent, took place, and the party proceeded to the count-house, which they found to be, in fact, a miniature hotel, with all the necessary appurtenances, stables, servants, &c., and a long dining room, about 50 feet, handsomely fitted up with oak furniture, *en suite;* the walls hung with maps, and working drawings of the mine, showing all the various lodes, with their dips, levels, cross-cuts, &c. By the way, every count-house should be similarly decorated.

The various plans, sections, &c., having been explained by Capt. P., to the infinite pleasure of the visitors, whilst Capt. R. was reading the morning letters, the door was opened by the latter, who proudly said, "This, gentlemen, is the place in which we declare our dividends." "And pray, captain, what amount of dividends may you have had the pleasure of declaring on this mine?" "Oh, I cannot tell to a few thousands, but I suppose about two hundred and thirty or forty thousand pounds, and we shall sample about 5000*l.* to-day, I think; shall we not, Capt. P.?" "Yes, Sir, about that." "Besides this dividend," continued Capt. R., "we are content to lose 600*l.* every two months in that mine;" pointing to one on the side of the hill; "but it suits our purpose."

"Is it possible!" was the response, "and what is your capital?" "Why, we are out about five guineas on each five hundred and twelfth share. Take these gentlemen into the office, and let them see the dividend account, Capt. P." On returning, they loudly expressed their surprise and delight; they were then asked to take a glass of wine and biscuit, before setting off on their day's journey. This done as quickly as possible, they adjourned to the floors, where the processes of bucking, jigging, buddling, &c., were witnessed, as well as the various modes of tin dressing, with which, having seen them before at Great Hewas, they were tolerably conversant. Thence they were conducted to a busy scene indeed. There were the great, the important, men of the day—the "samplers"—on whose report so much depends, superintending the turning, re-turning, cutting and re-cutting, cross-cutting and re-cross-cutting, the numerous "doles," as the parcels of ore are termed by the tributers and miners, who on this day (to them an important one) work vigorously, and with a right good will, until the piles are thoroughly mixed. The samplers then proceed to fill small sample bags, containing 4 or 6 ozs., from various parts of each pile, which they take away with them to ascertain the percentage of copper contained in each lot, so as to tender for it on the "ticketing" day.

As may be supposed, sampling-day on the mine is one of considerable bustle, the samplers being generally in haste to be off to another mine. All hands are employed as busily as bees to get the job done before dinner. To persons who have never been on a mine on sampling-day we would advise them to do so, as on a large mine like Wheal B——t the sight is worth seeing. The tributers rally the samplers on the last shocking bad prices; and they, in their turn, joke the tributers that they don't know how to dress their ores better. It is a day of glee generally; a little latitude is allowed, not much work being done underground on that day.

After an introduction to, and a little conversation with, the samplers on their labours and duties, our party are reminded by Capt. R. it is time to be off to the B——r Mine, as it is half a mile or more to it, and three miles further to T——n. "Never mind, captain, we will soon manage that, as we have taken the precaution to order the carriages to come up for us."

At Wheal B——r their surprise was raised to wonder at the magnificence of the plant and order of everything about it; but still more so when they learned that the profits on the second working amounted to 200,000l.! Their time being limited, after taking a hurried view of this extensive property, much against their will they started for the T—n Mine; arrived on which they went to the count-house, a still more handsome one than the B——t's, and were electrified to hear this mine had divided 500,000l., on an outlay of little more than 3000l. The man-engine next attracted their attention, it having been explained and demonstrated by the ascent and descent of the miners. Capt. R. jocosely remarked, "As soon as you are ready, gentlemen, we will change and go down, if you please, to the 300. I'm going, and remember your promise." However, the visitors back out of the polite invitation, as there was not time (?) and repaired to the great engine (80 in.); admired the beauty and precision of the immense machine, and were astonished to find an accident had occurred here some years since, when the enormous "bob" (main beam) suddenly snapped in two; the piston-rod knocked out the cylinder bottom, and nearly destroyed the magnificent engine. Having taken a hasty surface view of this mine and the Gwennap Consols, United, and other celebrated mines being pointed out and explained to them, the party returned to Wheal B——t, delighted to have been on such a mine as the T——n, and to know it is a dividend mine yet, though having already returned such magnificent profits.

On their arrival they were received by Capt. P., who announced dinner as having been waiting an hour or more. Our party were ushered into the dining room, the tables of which were nicely laid out for about 30. A fine piece of roast beef, leg of mutton, &c., were soon on the board. The company, consisting of the principal agents of the mine and samplers, took their seats according to rank, Capt. R. occupying the head of the table and Capt. P. the bottom. Grace having been said, and the first course removed, our party were invited to take wine. Champagne was placed at the head of the table; a plain rice pudding, followed by bread and cheese, completed the repast. The captain then asked, generally, "Shall we have a drop of custom," which our party little supposed meant neat brandy, until they saw the bottle put upon the table. One of our party objected to the *aqua vitæ*, when he was twitted on his promise by the captain. "You promised to become a miner to-day, and you won't find any of them object to a glass of brandy on sampling day; will he, Capt. P.?" This joke elicited a titter all round, and our visitor complied, on Capt. P. saying, "I should think not, if I judge from myself." The cloth drawn, "The Queen," "Duke of Cornwall," &c. having been duly honoured, the toast, "Success to Wheal B——t, and better prices for our produce," was drunk, and was replied to by the elder sampler, who said nothing gave them more pleasure than to bring news of better prices to such a mine as Wheal B——t. That, however, did not lie with them but with their employers: they had only to give the quantities and qualities of the ores, and, judging from appearances, they thought they never had a better sampling on the mine. A magnificent bowl of punch was now produced by Capt. P., who had been brewing, and placed before Capt. R., who remarked, "Perhaps, gentlemen (I mean you pedestrians), you have not tasted count-house punch, and you shall drink the healths of the lords of the manor." The visitor previously objecting assured the captain he never took spirituous liquors; but in vain, the captain was soon

down upon him, with "Remember your promise, and even teetotal-
lors don't refuse this toast and count-house punch." He complied,
and found count-house punch so good that he never required a
second pressing that day as the glass went round. In half an hour
after dinner the agents, samplers, and all withdrew, took leave of
Capt. R. and friends by a hearty shake of the hand, and a warning
from the former to be sure and not "prill" the samples!

After their departure, our visitors requested to know if such treats
were provided by the liberality of the captain or at the expense of
the mine. On being answered the latter, of course, they stated their
astonishment, when the captain said, "Wait until you get 8*l.* or 10*l.*
a share dividend, every two months out of about 30*l.* outlay, and
you won't object to your agents having and giving a good dinner;
and by the time you have been mining as long as I have you will
learn a good dinner, good wine, and good company don't injure
a sample, but, if anything, improves it, you may depend." The
inuendo was understood and cavil silenced. "Besides," continued
the captain, "the lords of the manor send us a few dozen of champagne, and well they can afford it, receiving, as they do, thousands
a year dues from an estate that would scarcely graze a goose."

Capt. P. being requested to procure a few specimens of the ores
as mementos, soon brought in a tray filled with choice examples of
native, peacock, grey, yellow, and other copper and tin ores, which
were highly valued by our tourists, who accompanied the worthy
captain to his hospitable mansion, where the evening was spent in
recounting the adventures of the day; the captain going to the hotel
to see them home, and they to dream of the pleasures of sampling
day at Wheal B——t.

## THE FIRST DIVIDEND

THE LIFE OF A "TOAD UNDER A HARROW", OR A "COCK AT
the stake on Whitsuntide," is happiness itself, compared to the trials
which await the captain and secretary of a poor mine. No accident
occurs but it is their fault; no shareholder is in arrear but it is their
neglect; no unfavourable circumstance can turn up but they ought
to have foreseen it. The Chairman says they ought; the adventurers
take up the word, and echo they ought, and with faces as long as
fiddles cry out, "Away with them, away with them!" An old

Cornish adage on these matters says, "A good mine makes a good captain." This is partially true, but that a good mine makes kind and good tempered shareholders is a truism. How different is the outstretched arm, hearty shake by the hand, or graceful recognition, and "Well captain, who d'ye do? I see you've a capital floor of ore —good dividends next meeting, my boy!" to being stopped by the adventurer on the high road, who, without preliminary of any sort, grumbles out, "What the devil is the use of your boring me for that last call so often; when I choose to pay I will do so." "But, Sir, all say the same, and you know men won't half work, and take liberties, unless they are paid regularly. You little know what a life I lead with them when there is no money on pay-day." "I don't care a d—n about the men, or the mine either; you are all a set of knaves, and intend to ruin all that you are connected with."

No youth emerging from boyhood, or the constraint and supervision of his guardians, rejoices with such glee as captains and managers of mines on the declaration of their first dividend. All their cares, turmoils, troubles, taunts, and insults, are forgotten and forgiven; a little vanity or boast on their part must be pardoned. If they assert they all along knew the mine would early be in the dividend list, the fact of its being so should plead for them, and scrutiny wink at egotism.

Our friends, who had now almost identified themselves with mining, had entered fully into the exciting circumstances connected therewith, as much as if they had been accustomed to it for years, and were determined to see everything relating to the subject, witnessed that which is not an every day occurrence—the first dividend.

At a mine near Truro, on a late occasion, such was the case. The author also kindly received an invitation to be present, but having made arrangements for examining a deep mine, he was reluctantly obliged to forego the joyous scene; this being so recent, and our previous descriptions, relating to some years since, to alter the time of action would be a species of anachronism; therefore, we will confine ourselves to our previous period and former friends. The S—— Consols being the scene of the denouement, where the ceremony was carried out in very good style—in fact, arranged with that liberality and discretion such occasions require: having previously given full particulars of all proceedings, we may as well enter on the minutiae of this.

The mine had been twice before abandoned as worthless by

ignorant committees, who did not know what they were about, and would not be told, by these means getting into inextricable difficulties, which ended in quarrelling among themselves. It had been now resumed for the third time, the appearances of the lode warranting the opinion that it would ultimately be found profitable. Even by the present party it had been decided to call a meeting for giving notice to the landowner to discontinue working, but the agents always assured them that it was worse than madness to do so, as they had done so much work, which would, of course, be for the sole advantage of the landlord, who could with propriety demand a corresponding premium from the succeeding parties desirous of working the mine, which would not be idle a single week. They urged, begged, and entreated the meeting not to be guilty of such suicidal folly—such ruinous resolutions. Then professional opinions (without number or consideration of the antecedents, or experience of the inspectors) were consulted, the committee changed, the Chairman ousted, the shares fell in value to a tremendous discount, the captains got notice of dismissal, the mine and merchants' bills into arrears of pay; in fact, the whole affair, by the busy-body exertions of the "heutontimerumenos" (there is one in every mine, as there is one fool in every parish), had got into disrepute. Even the true shareholders began to despair, when lo! the cross-cut south revealed the secret they had so long been seeking. Corresponding cross-cuts were driven from all the levels, and the agent's original opinion found to be correct. The multitude of "reports" had bewildered the committee, who in one month ordered the 20 to be driven first east, then two days after to be stopped and driven west, then stopped, and the adit to be prosecuted with all speed, when as suddenly an order would come to suspend all proceedings on the mine, forgetting that salaries, rent, engine, and other expenses, must be kept up, whether the miners worked or not; and so they had gone on for months, ruining their own property by their own ignorance, to the disgust of the landlord and their own loss, squandering their money without any definite object. Whatever was wrong, the "heutontimerumenos" was sure always to be right, having the happy knack, as they all have, of foreseeing events after they have occurred. When this discovery was made he, of course, had predicted it. Be that as it may, the discovery was made, large sales were effected, the debts of the mine paid, the unaccommodating bankers held a handsome balance in their hands, and the secretary had the satisfaction at last of announcing in the circular calling the quarterly meeting that a dividend of 10*l*. per share would be proposed.

As had been expected, this was the best attended meeting ever held since the mine had been worked: few shareholders were absent, as they were all anxious to know how these things could so soon be, and be genuine. Many were the heart-burnings and regrets at having sold shares at ruinous prices, and many the joyous purchasers, not a few of whom were by the losers accused of having some secret information that had been withheld from them. There, too, was the busybody, to give every person information he did not himself possess, stating again he had all along foreseen that happy event, but had held his tongue; he had not been willing to hold out hopes that might have been delayed. The captains boldly declared that had it not been for his meddling and utter ignorance, by which the committee had been partially influenced, the mine would have been dividend-paying two years before. Preparations suitable to the important occasion had been made: the engine-house and other buildings on the mine had been whitewashed and painted, the roads fresh laid with waste from the buddles and jigging-machines, a new flag hoisted on the shears, the floors swept up, the doles of ore nicely rounded, and everything made as neat and spruce as possible.

The day of the meeting having arrived, the now reinstated Chairman, with the old committee and friends, in a carriage and four with postilions, drove up to the count-house, at the door of which they were received by the lord of the manor, the smiling parson, the doctor (both of whom had purchased shares at a discount, by the advice of the captain), the secretary, captain, and our tourists. As soon as the appointed hour arrived the Chairman took his seat, amid a buz of applause and subdued signs of gratification. The reports having been read, and, of course, received and adopted as highly satisfactory, and the accounts being duly audited and attested, were passed over without that scrutiny and squabbling heretofore inseparable from such occasions. Even the grumblers did not find out they had been cheated that month out of half a candle or two-penny worth of useless chips; when the Chairman rose and said, "Gentlemen, it now becomes my pleasing duty, and, believe me, I never had a more gratifying one in my life, to ask you to take into consideration the propriety of declaring our first dividend on the Wheal S—— Consols. (Hear, hear.) It appears from the accounts just read that we have in our bankers' hands assets to the amount of 2800*l*.; our next month's cost-sheet will be about 400*l*. The captain says the ore you have this day seen on the floors is worth from 1000*l*.

to 1200*l.* more; this will be sold in a fortnight from this time. All our new machinery is paid for; in fact, the mine is perfectly out of debt save the current month's expenses, the reserves are good, our daily explorations are satisfactory, and are such as to assure us that our patience and perseverance will be rewarded by a great and lasting mine." Turning to the captain—"Capt. W——, be kind enough to leave the room for a few minutes with Mr. G——, the secretary. Gentlemen," addressing the meeting again, "the committee beg now to recommend that a sum of 1000*l.*, being 10*l.* per share on the shares of the S—— Consols Mines, be now voted as a dividend, to be payable on the first day of July next; and, further, that the sum of 100*l.* each be presented to the captain and secretary as an acknowlegement of their services, ability, perseverance, and encouragement they have always shown to the mine and adventurers." (Hear, hear, and considerable applause.)

This was seconded by the lord of the manor, who, being resident on the spot, could testify to the zeal, attention, and solicitous care displayed by these officials, the only dissenting party being the "heutontimerumenos," who asserted, if any reward were to be given, he ought to have it. The mine would have gone down had it not been for him; he had been its main stay. He had exerted himself greatly. He had only 10*l.* granted to him and though he had not obtained the situation he intended, yet he had saved the mine. He had kept the captains in their situation and the secretary's services (he was the pink of a secretary) had been secured by him; therefore, he did not see any policy in throwing away money in that manner. He held 100th part of the mine. It was of serious consequence to him, and he (nodding significantly) knew it was to others; he, therefore, begged to propose the whole sum be divided equally among the shareholders. Not finding a seconder, however, the dividend and gratuity were declared duly made (the grumbler remarking "Wrong again.") The secretary was ordered to draw out the cheques; the captain and he, with hearts too full for utterance, gesticulated and stammered out their thanks. A vote of confidence to the committee and Chairman was given amid acclamation, and the meeting dissolved.

Champagne and other wines were now produced. "Success to the mine," "Increased dividends," "Fish, tin, and copper," and other mining toasts, having been drunk (the personal ones being reserved till evening), the company spent an hour or two in examining the surface of the mine, inspecting the new dressing machinery, which had so much conduced to the day's dividend, and in visiting the temporary marquee, in which the miners were enjoying a good old English dinner, provided by a subscription amongst the principal adventurers on this auspicious occasion; and every person, not even excepting the grumbler himself, felt satisfied that day.

This having been accomplished, the whole party adjourned to the Royal Hotel, Truro, where a sumptuous repast was served, at 6 o'clock p.m. to the adventurers and invited guests; covers were laid for about 60. After the removal of the cloth, the usual ceremonies were gone through, until it came again to "Success to Wheal S—— Consols," before proposing which the courteous secretary handed in sundry slips of paper, on which were partly written, partly printed, "Please to pay So and So, Esq., the sum of so much, on account of the Wheal S—— Consols Mines." Many were the pleasant jokes that passed respecting these bits of paper, and the marked improvement in the secretary's handwriting; many had never seen it look so well before, or so easily understood, and many a smart repartee was the secretary enabled to give those persons who had repulsively treated him when he was absolutely doing his duty, and preparing the way for realising this happy event.

The evening was spent, as it may be supposed all such evenings to be and are spent, in kindly feeling towards each other, which waxed warmer in expression as the hours approached towards midnight, and in the toasting of captains and agents, committee and chairman, landlord, and visitors. On this occasion we believe they were honest; even the grumbler was in his best suit and best temper, as he absolutely laughed and cracked a joke.

And now for the moral. Oh, Chairman, committeemen, adventurers, and grumblers! depend on it this is a true picture, painted from and by Nature itself. Therefore, if your secretary be urgent, be not unkind. If your captain be able, though sanguine, do not discourage him; if not equal to his duties, discharge him, but do not mystify him or yourselves by a multitude of opinions, one-half of which are given by pretending nincompoops. When you have an able Chairman, keep him. Nothing tends to shake public confidence in any undertaking so much as continually changing the executive; by the time they have been inaugurated, and know their duties, their term of office expires. Above all, do not commit the management of mines to costermongers, barbers' clerks, or ignoramusses, and you too, like our pedestrians, shall assuredly have the honour of attending the "first dividend," and that on your own mine. This

picture is so common that it may be taken as a general portrait. All
the picture is true, and parts of it will be recognised.


MINE COUNT-HOUSES IN 1827, AND MINE COUNT-HOUSES
IN 1857

NO ORGANIC CHANGES IN THE HABITS OF ANY SPHERE OF
society can take place without more or less influencing other circles;
these act in degrees as they come in contact with the good or evil.
Perhaps in no instance is this more apparent than that to which we
allude in our title. Education, and the more refined taste now being
so much cultivated, have extended to even the remote districts
which are the scenes of our previous descriptions in these papers. If
any causes were pardonable for the excesses sometimes indulged in,
the actors had certainly valid excuses—viz., the excitement always
created by wealth accumulated without labour, in many instances
without outlay of capital, an almost uncontrolled exchequer, a want
of education, and more than all these combined, examples and
precedents among their patrons and employers.

To be as drunk as a lord was at that period not deemed a venality;
even the highest personage in the realm and his associates, though
men of extraordinary and known talents, boasted of being four-
bottle men, and to be "well seasoned" was deemed a gentlemanly
qualification. They who aspired to and frequently possessed wealth
naturally followed in their wake: hence a considerable amount of
the extravagance then practised is now condemned; though almost
extinct, the memory still lives, and has a baneful effect on the really
necessary conveniences and conventionalities of society. These
evils were, however, carried to such excess that, were they even
attempted to be revived, they would not be tolerated for an hour;
indeed, it would be difficult to find participators.

At and before the first period alluded to, a mine account was
generally held on the mine itself—a very good regulation, where the
adventure is local; after which a dinner was provided for such of the
proprietors as chose to attend, at the expense of the mine, also a fair
and good regulation, as they who attend the business certainly
deserve some kind of compliment; besides, the mines are frequently
so remotely situated as to render refreshment for their horses as well
as themselves absolutely necessary: hence the requirement of stables

and horses on a mine—items so frequently objected to by distant committees. These expenses, however, are very different from the extravagancies to which such things were carried formerly. Mining count-houses were more like regular hotels than mere places for transacting mercantile pursuits. The bottle was to be seen on the table throughout the day; the dinner most frequently open to persons who really had no business whatever on the mine but an excuse to get a "blow out," with wine, punch, &c., *ad libitum*, and a drop of "warm with" to go home in the evening. The going home generally meant an adjournment to the nearest village public-house, the captain or agent, of course, bearing company, where the scenes of the day meet with a proper finish. These practices not only entailed considerable expenses on the mine, but interfered sadly with the captains and agents' time and attention, besides bringing them into loose habits. Be that as it may, it all passed off as a matter of course. Captains, agents, adventurers, merchants, pursers, samplers, clerks, and diners out, were "All hail good fellows well met," and so the jovial glass went round at the owners' account. There positively have been instances in which a call to liquidate the dinner cost has been as necessary as to work the mine; this may appear almost incredible, but it is no less true. The mines at that time, except in a few instances, were divided into no more than 56 or 112 shares each, all of them probably held in the immediate neighbourhood of the mine. Perhaps 40 or 50 of the shareholders would be present, and as, according to custom, the charge was made to the mine, and the more the merrier, the amount soon swelled to a considerable sum.

We have known instances in which such extreme disgust has been excited by similar proceedings as to deter right-minded persons from joining a speculation altogether. One came specially under our notice, where some managers of the old school were about commencing a mine. The first necessary step was deemed a "glorious spread:" parties who had joined, and parties who it was hoped and expected would join the scheme, were alike invited, music engaged, flags floated in the breeze, the miners were regaled, and a grand field day prepared; a sumptuous dinner was provided, wine flowed like water, and the good things of this world dispensed with a liberal hand. Many wealthy persons were present, among them a wily London banker, of fabulous property, who, when taking his leave, shrewdly desired to know how this feast was to be paid for. On being told by the fussy little secretary it would be charged to the cost of the mine, he at once said—"Then cancel my interest, or I will give away my shares, as I will suffer no set of men to guzzle my property in this manner. I thought and expected that every gentleman in this room would, of course, pay his quota, and not partake of such expensive repasts at the cost of his absent fellow-shareholders, or permit others to do so, when not on official business of the mine. No, no! I've done; I'm content with my first loss." Thus was this would-have-been valuable acquisition for ever lost to mining; nothing on earth would induce him again to attend a preliminary meeting on a mine.

These were the palmy days of mine establishments. The terrible crisis of 1825 and 1826 shook mining to its very foundation. Previously to these years one of the periodical manias for such speculations had set in, and continued for some three or four years without abatement or check. All was *coleur de rose;* everybody was to be as rich as Croesus from mine produce; everything went on as merrily as a marriage bell, or as Mr. Hudson in a rising share market; there were then no losers. The mine that made the greatest display at the surface, and produced the best count-house, was most talked about as to what it "would eventually be," and, as a matter of course, was most enquired after for investment. The facilities for correct information, be it remembered, were not then as now; there was no *Mining Journal* to chronicle facts; the localities were seldom visited; the mines surveyed or reported on; the books examined by auditors or the outside shareholders. At the meetings the bills and statement of accounts passed *nem. con.,* and so the meetings ended. The financial accounts under such circumstances, as might have been expected, were little better than inexplicable jumbles of figures—a hotch-potch leading to error, and opening the way to peculation and fraud. At the payment of merchants' and tradesmen's bills, too, a bonus, termed "hat money", was frequently handed over to the purser or captain, as well as presents periodically given. These customs have also been nearly exterminated, the patronage being vested more in committees, or destroyed by the system of tendering, competition being now very great. It is still practised, however, in some few mines, but is gradually declining, and will shortly become extinct.

In 1827 and 1828 came the crash, the result of the commercial panic which had occurred a year or two before, ruining many of the proprietors. The bankruptcies which followed made sad havoc with mining. Reckless expenditure and deceptive schemes were in many instances exposed, and in many, sadly too many, were the faults of fraudulent merchants palmed off as the losses incurred in mining, whilst mining had no more to do with their defalcations than they had with the loss of America. Have we not seen the same thing over and over again? How many ruined spendthrifts have ascribed their difficulties to railway speculation, who never bought or sold a share? How many who never had a shilling of their own to lose, after setting up as "stags", say they have lost their all in speculations? This is but the history of mining (so called) over again. The effects of this species of misrepresentation has been felt to this very day. Many, doubtless, gambled in the mines, their only aim being to realise money for themselves out of the cupidity or credulity of their victims. Hence, and to suit such purposes, these schemes and extravagancies were sanctioned, and even fostered.

The reaction having fully come, the mines that were not remunerative stopped, almost without an exception. Good, bad, and indifferent were all included in one category, were stigmatised as rank swindles, and the money subscribed spent in eating and drinking. Condemnation was as rife as exuberant praise had been before.

Many a mine thus recklessly abandoned has been re-worked, and returned its hundreds of thousands of pounds profit, and many are still un-wrought. Evil frequently begets good of itself; "out of the eater came forth meat." These circumstances, if not the cause altogether, were in a great degree the forerunners, the harbingers, of a better state of things, which the rules and requirements of society render absolutely necessary. No such revels can now take place even in the most liberally provided and richest establishments. Except it be on occasions such as sampling days, pay days, and the like, seldom are there any expenses charged to the adventurers: of these we have given true and faithful descriptions in our previous photographs, in which we witness only liberal hospitality, or polite courtesy, presenting a strong contrast in the periods indicated in our title.

Count-houses are now what they were originally intended to and ought to be—viz., proper offices for keeping the mine accounts for the meetings of the adventurers, for the reception of persons on business, for the temporary residence and sleeping apartments of the captain when night captains are employed, or in cases of emergency, in situations at a distance from the captain's house.

In all well regulated mines, a certain small sum is allowed monthly for the count-house expenses; these are absolutely necessary; there should be a medium drawn between excessive expenditure and extreme parsimony. Many mere huts, constructed of turf, or a few boards nailed together, neither wind nor water-tight, a disgrace to the proprietary and discomfort to the captain, are dignified by the name, unsafe alike to the security of the documents necessary to be therein kept and preserved, and to the sums there deposited: this is assuredly a false economy. Where no such building exists, a neighbouring public-house is resorted to as a substitute, at which the men are paid, and business with the various merchants transacted, the evils of which are so apparent as to need no comment.

Let any person visit the counting-house of a well-ordered mine in 1857, and he will find an efficient staff, neatness, order, comfort, and regularity, united with economy and convenience. This has entirely resulted from the new order of things: the principal proprietors have set a good example, the which is emulated by all connected. Where a moderate sum is allowed, the rule is seldom broken or exceeded, but where a false economy refuses such trifles altogether, necessity drives them to temptation, and irregularity sometimes occurs; it is, therefore, wisdom to be moderate in these as in all other things, for it may be relied on, none of the scenes formerly so common, and so justly complained of, are extant now; and that there is as marked an improvement in the keeping of accounts, and in the observances and customs of count-houses, as there is in the customs and conventionalities in everyday life of 1827 and 1857.


## THE MINERS' HOLIDAY—MIDSUMMER-DAY

IN FEW PARTS OF ENGLAND ARE THE REMEMBRANCES OF OLDEN time so cherished and regularly observed as among the people of Devon and Cornwall. In these counties the feasts, fairs, and customs existing from time immemorial are kept up, though shorn, as all such antiquated affairs are, of much of their pristine consequence and celebrity. Amongst them, the most conspicuous is the Flora or fairy dance, at Helston, on May 8: it is still patronised and participated in by the neighbouring and resident gentry. It is said by antiquarians to be the anniversary of a festival of Flora, which goddess is supposed to have had a temple near or in this place; little doubt seems to be entertained on the subject. A ridiculous story or legend is related of a fiery dragon flying over the town on some remote occasion, and the rites attributed to its commemoration. Be that as it may, the Helston people have, and celebrate their feast, which partakes more of the old Bacchanalian characteristics than most ceremonies of the kind. Gaily dressed in their best attire, and decorated with a profusion of flowers, the performers, preceded by bands of music, dance through the principal streets, and through the larger residences in the town, in the early part of the day, which performance is again repeated in the evening. The music is a peculiar light, merry air, known as the Flora dance. No work is allowed to be followed in the town on that day, a heavy punishment awaiting the transgressor. The day is given up to a sort of *abandon* to the dance, music, and innocent revelry, in which all grades of society, ages, and sexes are participators. It is not even deemed degrading to civic dignity to honour the balls by patronage, the chief magistrate often being the gentleman to lead off at the principal hotel.

The fishermen, too, have their holiday; St. Peter's day is their festival, he being their titular saint. This celebration may be easily accounted for, and appears not so much out of keeping. At Newlyn (west), near Penzance, and a few other fishing villages, it is still tolerably well honoured, but is fast falling into disuetude. As soon as darkness approaches, a number of young people run through the streets, swinging huge blazing torches round their heads, and persons, even women and girls, may be seen practising the apparently dangerous pastime; tar barrels blaze from every prominent position, fire-works of every description, but of very inferior quality, are let off in abundance. These orgies are kept up until midnight, when a long string of youth of both sexes pursue their way through the streets, amusing themselves at the game of "Thread the needle."

Such being the customs in vogue among the people generally, it is not surprising that the miners—who, by the way it should be observed, are looked upon by all persons and regard themselves as a peculiar class, seldom mixing or associating with farm labourers and mechanics—should have their holiday. All work and no play is not their motto, or their practice. Their own "day" is Midsummer-day, which is held by them in high estimation. Whence the custom originated is lost in oblivion, but it is conjectured to have been, like the Flora, the remnant of a Pagan ceremonial, as within the memory of many now living it was the day on which wrestling, a game in the West of England, confined to Devon and Cornwall, was universally practised in these districts; this game has been played from time immemorial by miners and tinners; though still much encouraged, it is far from being so much in vogue as it was even some 20 or 30 years ago. The late Dr. Paris ascribes these games on

Midsummer-day to a commemoration of a feast of Hercules, or some other heathen divinity, introduced by the Romans or Phœnicians, both of whom are known to have traded to these parts for tin; what seems in the doctor's opinion to strengthen the idea is, that the paintings and statues of combatants of the atheletæ at the Olympic games, still extant, correspond materially in their attitudes and actions with the wrestlers of these counties at this day. At present, wrestlings are frequently held on the miners' holiday, but not to any very great extent.

On most mines, the men on this day whitewash and clear out their changing-houses, the outside of the mine buildings, &c. Where the adventurers will afford it, great pride is taken in having the building and woodwork re-painted, the dressing-floors swept up, and everything put into perfect order. If the mine be productive large branches of green oak, called the "Midsummer branch," are placed on the shears of the mine, and a new and gaudy flag is attached to the vane, containing the initial letters of the name of the mine, and fixed on the highest part—the shears, if there be one, being usually the place selected. Little or no work is done underground on this day, it being literally and veritably a holiday. The men are always allowed one shilling, and the women, boys, and girls sixpence each, as spending money. By mine proprietors this is seldom objected to; indeed, it is one of those old customs it would not be well to attempt to abrogate. Surely, one shilling per man per annum cannot do much injury to a mining company. The greatest mischief is done by the men going to the public-house, where not only is the shilling spent, but many a hardly-earned one is added; there is, however, much less drunkenness prevalent than was wont. The change of manners, effected by various causes, is nowhere more visible than in our miners, who were formerly notorious for their spendthrift, dissolute habits, rough demeanour, and uncouth language; still, there is the old leaven left, which occasionally appears, to use their own apothegm, "when the liquor is in the wit is out." On the whole, however, their progress in society is most encouraging; they are now noticed as a particularly intelligent, shrewd, civil, and thrifty set of men. The Cornish *patois* is fast disappearing; many words which ten or twelve years since were in use are now obsolete, except in the most remote places.

But we digress, therefore, let us return to the miners' holiday. The Midsummer-day is now generally spent in parties of pleasure, and excursions to the neighbouring towns and villages on the sea coast. The opening of the railway to Penzance and Truro has afforded facilities to the vast mining population of the Camborne and Redruth districts, of which they have not been slow to avail themselves—thousands enjoying a cheap trip to the former watering place to get a "pennoth of sea," and to the latter to view the "lions of Truro." As soon as the railway shall have been opened to Plymouth, further excellent opportunities will be afforded to gratify that innate desire all feel to gain information by experience and going from home; even though it be in so limited a degree, it is an evident improvement on the old system. Now, the question presents itself, Would it not be well for the proprietors of large mines, or by union among small ones, to organise and patronise such trips? No doubt the men as well as the children would willingly spend their Midsummer money in such amusements, in preference to loitering about home in their own neighbourhood: there would be more really useful lessons and morality taught by such agencies than by the most laboured eloquence or devout exhortation. There might also be a great improvement made in the amusements provided at Penzance, and other watering places along the coasts of Devon and Cornwall on this day, which is also a general holiday—the exhibition of Manby's and other apparatus for saving life in cases of shipwreck, and the use of the life-boat exemplified, by which means our working population in these localities would become familiarised with the uses of these invaluable discoveries. A small premium would secure attention and create emulation. The officers of the Coast Guard could and gladly would experiment and instruct. By these means much good might be accomplished, much useful knowledge conveyed, and that, too, through the medium of pleasure, which is far more easily effected and better relished than by any other method. This practice may be useful to all classes, as even the wealthier would be attracted to the scene; and who knows that even they may not require such knowledge?

Had these displays been as familiar as they should have been on a coast like that to which we refer, or even throughout the whole island coasts (for we are essentially a maritime people), many a harrowing detail of shipwreck would have been spared. How frequently do we read of apparatus having been on the shore in the time of distress, when no one knew how to use them. And how often has the rope, even when thrown on board the vessel, been misapplied, by being fastened to one man, and drawing him through the surf, instead of hauling a rope on board the stranded vessel, and by his ignorance losing probably his own life, and depriving his fellow-shipmates of the messenger of hope and of deliverance? From their familiarity with danger, their daring habits, their knowledge of the uses and powers of gunpowder, few men are so well adapted to the pursuits as our miners, from the circumstance of many of the mines being situated in places where these casualties occur; and from the numerous opportunities they have of boating in the various creeks and harbours, in some parishes—for instance, St. Just or St. Agnes—many of these men are half sailors, and from their habit of continually emigrating to all parts of the world, they are pre-eminently the class of men to whom such knowledge should be imparted, not only for their own personal benefit, but for the good of mankind. It is said every man should be taught to swim; true, it should be so. This practice we agree is quite as necessary, and would be of great benefit. The men and the places are available; it only wants an example to be set by some benevolent and influential individuals as an experiment, which we are sure would be successful and highly attractive, would be a great benefit, and would convert the miners' holiday, Midsummer-day, from a scene of uselessly expended vigour, energy, and strength, of vice and drunkenness, into a day of cheerful recreation and instruction, to knowledge of really great benefit to themselves, and to a great improvement morally, intellectually, and physically of our industrious population. We offer these practical suggestions to those whose duty it is to see to the proper application of the "Miners' Holiday, Midsummer-day."

## THE CHAPEL

NOTHING STRIKES STRANGERS ON THEIR VISIT TO THE MINING districts of the West of England more forcibly than the great number of dissenting chapels. In every town several are to be found of all denominations—save unitarians and Roman Catholics, few communicants of these creeds being residents. In every village and hamlet, however remote, at least one or two of the various offshoots of methodism are always to be seen; even on the highways or common downs, where the roads converge from a few scattered cottages, there, although not a house is to be seen sometimes for miles, is sure to be a chapel of some sort, however humble, frequently consisting of four mud walls and a little thatch, sometimes the rude work of the miners and labourers themselves, and are chiefly of the methodist or baptist denominations. A variety of causes have led to this general practice of chapel building. The increase of population that has converted hamlets into villages and villages into towns has, undoubtedly, been one cause. There can, however, be no doubt whatever a great deal has been owing to the laxity of the clergy before and at the time of Wesley's journeying in these parts, which indeed became one of the principal scenes of his labours, and where his name is still held in universal and profound respect. Even within the last 30 or 40 years the parochial duties were disgracefully conducted, or altogether omitted (now happily completely reversed), particularly in the very districts in which active ministry was most required. These were, nevertheless, far from being the whole of the reasons—these were only two amongst the rest. In very many instances the church, instead of being placed in the midst of the parish, is in some remote corner or out of the way place, as at Mylor, St. Just in Roseland, and many others. In some places the majority of the parishioners had to walk or ride four or five miles over a dreary moor or a wild common. At Brent Tor, in Devon, the church is absolutely perched on the apex of a mountain, a most picturesque but most impracticable situation, certainly. Such a journey, in so wet a climate as there prevails, often acted as a drawback to due attendance on divine service. Besides, the pews in the various churches were claimed, whether occupied or not, by the proprietors of certain estates in the parishes who paid tythe, and it is not too much to say the spiritual welfare of the mining population was entirely and grossly neglected. The agricultural labourers, on the contrary, were provided with a pew or seat by virtue of the farm tenure, and were duly expected by their employers to appear at church once every Sunday at least, if service were performed.

In many parishes, in which the livings were united in twos or even threes, the church prayers were never read at all in some of them. Though the church itself existed as a building, there was literally no service, and no congregation as a matter of course, though there was a parson to be paid and souls to be saved. The stranger will be struck with the number and beauty of the towers and ecclesiastical structures of great antiquity to be found in these districts, showing the piety of the builders, and their evident preparation for, and anticipation of, larger congregations.

To so great a neglect had the matters arrived, that to this day there is a standing taunt in the parish of Morvah, in Penwith, where it is said a "Cow ate the bell-rope," the church having fallen into such utter decay that the tower-door was gone and the bell-rope rotten. A farmer of the parish dying, it was necessary to toll the bell at his funeral; a hay-band being substituted for the rope, a hungry cow straying about wandered into the churchyard, and discovered the dainty morsel; in enjoying the repast, she tolled the bell, to the no small dismay of the occupants of the adjacent public-house (the only residence near), who attributed the awful midnight irregular clamour to a ghost, or some unearthly visitor. The story is said to be perfectly true, and is very probable. At St. Ives the church had sunk into such disuse and neglect that the splendidly decorated stalls were actually destroyed, the beautifully carved oak roof whitewashed, and the carving obliterated! The very foundation of the fabric itself endangered by the sale of vaults and graves! whilst divine service was only occasionally performed in a population of upwards of 4000. In the parish of Stythians, near Penryn, in which hundreds of miners and granite workers resided, service was only observed once in two or three weeks, and then at such hours as were very inconvenient to persons in humble life. Even where services were regularly performed, except in large towns, the clergymen, though undoubtedly sincere, and some of the best-intentioned men, were, in many instances, ill-adapted for, and by nature physically incapable to fill so important an office as pastor of a large parish, whilst the injudicious selection of parish clerks tended, in no small degree, to render that which was already bad still worse, as instanced in the parishes of Budock and Mylor, near Falmouth, the officiating ministers in each of which were as devout, as sincere, and as zealous men as could be desired; but from defects in articulation were scarcely to be understood, except by persons accustomed to their delivery, whilst the broad Cornish sing-song nasal reading of the clerks, rendered the sublime to border on the ridiculous. We should by no means advert to these circumstances, but to show the utter neglect into which public religious duties had fallen at a time so recent as 30 or 40 years since. The natural consequences resulted—few or none of the working population visited any kind of place of worship, but sought some retired place—even sometimes close to the village, and publicly—the men to wrestle or play at pitch and toss, and the boys to enjoy a game of marbles, or other idle pursuit, in which they were frequently joined by their elders. No idea of neatness, cleanliness, or self-respect was cherished; they were then literally "the unwashed;" an amount of ignorance and superstition was prevalent which at this day is barely possible to be conceived. The clergy had not, and sought not to have, any hold on these people; they certainly were a rough, uncouth set of beings; to reason with them on the sublime precepts and gentle teachings of the gospel would have been next to useless. Whitfield, Wesley, and their early zealous coadjutors, who so energetically appealed to their sympathies, and wrought on their fears in their plain but eloquent discourses, were the very messengers for these totally neglected individuals. If the populations would not come to them, they, despite every difficulty of prejudice and persecution, sought out and made their flocks, literally carrying out the command to "go into the highways to bid to the marriage feast." With the suddenness of magic the good seed took root, and has verily grown into a great tree. The number, size, and handsome character of their conventicles now bespeak a consequence and meaning far more important than are apparent to the mere casual observer. The services therein are conducted by a regular body of established preachers, and what are termed local preachers, many of whom are the captains or agents of the mines. Though some are certainly of dubious ability for such offices, yet there is a nervousness and sincerity of manner—a mode of expression—quite necessary for the work.

Their hours of worship, having earlier morning and later evening services, the much better situations and greater number of their chapels, have secured congregations far more numerous than those who attend the established church. This ancient institution must, at the present day, be most decidedly and emphatically exonerated from any charge of listlessness or want of exertion; but how very different now is the class of pastors having charge of these parishes, and how different the discipline to what they were when Wesleyanism and dissent reared their heads. The clergy are now distinguished for their attention, earnestness and desire to cultivate the spiritual and temporal welfare of the people; most admirable instances are to be found of undoubted piety and fervid zeal. Few of the old style before described remain; these as they die off are replaced by the new and far more effective race, and the improvement is everywhere palpable.

In the present day every church and chapel has its Sunday-school—a fact that incontestably had its origin in the chapel: that this early induction to something like discipline has been of incalculable benefit to the present and rising generations is admitted and imitated by all denominations.

Since we have previously introduced our friends to the society of miners in their places of business, amusements, and other scenes, let them accompany them to their village chapel on a Sunday morning, and they will see a number of neatly-clad orderly persons wending their way to the meeting-house, at the door of which, ere the arrival of the preacher, they will find a large congregation, among whom will be noticed some persons of superior appearance, usually the mine agents or respectable yeomen of the neighbourhood. Morning service is usually conducted by the local, and the evening by the district preachers, or *vice versa*. In either case a decorum, attention, and reverential manner is observable; the singing is of a superior character; the fine deep-toned voices of these stalwart fellows produce an admirable effect. Hymn singing is much practised by all classes of miners, and is carefully studied. At the dismissal the people retire in good order to their respective homes. The afternoon is usually spent in reading or taking a saunter in the country. The evening is marked by crowds thronging every road and pathway leading to the meeting-house, the majority being females, whose household duties had precluded visiting the morning service. The visitor will not fail to be struck with the particularly tidy, not to say smart appearance made by persons in so humble a sphere of life, presenting, as it does, a most powerful and favourable contrast to byegone times.

In no part of the kingdom is the voluntary principle more generously supported, their means being taken into consideration. The people, out of their hardly-earned wages, contribute liberally to subscriptions of all kinds, particularly chapel extension and missionary enterprise. Formerly landed proprietors objected to have chapels erected on their estates. The benefits accruing to the localities have greatly altered their feelings in this respect; now not only is the land usually presented, or leased at a nominal rent, but the proprietor frequently adds a handsome donation.

If we contrast the ancient and modern pictures, and both are correctly drawn, neither over-coloured or exaggerated, we shall admit a vast change for the good has been achieved, and is still gradually in progress, as the records of their courts of assize fully testify, the higher class of crimes being very rare, and, when they are detected, more frequently committed by strangers than natives.

If any rivalry between different sections of religionists exist it is in good spirit: with them, however, we have nothing to do; it is only our duty to portray as we find them, and to mark the social progress. This we have endeavoured to do to the best of our ability. Where the efforts of all are equally deserving, it would be invidious to particularise, especially in matters of so sacred a character. We trust the good work will prosper, and effect the glorious ends desired, by whatever agency produced. Still we cannot but confess the counties of Devon and Cornwall, the mining population most especially, owe much of their moral improvement, their domestic comfort, and their present condition in the scale of society and happiness to the "Chapel."

## THE CAPTAIN

IN THESE PAPERS, HAVING PREVIOUSLY GIVEN A TOLERABLY numerous and varied series of pictures of mines and mining customs, we will, according to our "prospectus," now endeavour to portray the officers connected with these important undertakings; in doing so, however, we premise that in such descriptions we necessarily state as much what they should be as what they really are, as all cannot be supposed to come up to the ideal standard of perfection— this we know to be impossible; indeed, should such *rarae aves* be found, we question if they would all have due credit for their abilities or endeavours. Where there are so many masters and interests to be served, opinions are so various that it is impossible to please all, even if all be well served. Captains, therefore, should carefully study the well-known fable of the "Old Man and his Ass;" they will find the straightforward paths of duty lead to the goal of prosperity far more quickly and directly that taking advice of any or everybody who volunteers to teach them what they ought to do.

As his complimentary title would imply, the greatest personage on the mine is the captain; he should, of course, be No. 1. Like his prototype on shipboard should his status be on the mine; the whole government thereof should devolve on him, and not, as is too frequently the case, on the dictation of a browbeating, overbearing chairman or committee, by whom he is thwarted in his endeavours, or rendered a mere cypher, endangering the remark once wittily made at Wheal Henry and Wheal Music (St. Agnes) by a mine inspector, who, after his examination of the mine, on being asked his opinion, replied—"I see you are famously off for captains, as you have no less than a hundred on the two mines; more captains than men."—"How do you make that appear?" enquired the adventurer.

The answer was, "I will show you on paper that it is a fact. Take your pen and write down Captain Oates as 1, and then let the other two captains be added as 00, which they really are, and you at once have 100, as they appear in these mines."

How this title came to be adopted in mine management is as much a matter of speculation as the prefix of Huel, or Wheal, to Cornish mines; its elucidation would be of just as much advantage in the one case as the other, therefore we shall not waste our time seeking its derivation, the parallel instance having puzzled many a clever brain to no use in the matter. Use renders all things familiar and perfect, and the term answers the purpose. "Captain" is certainly more euphonious than "white jacket" (the insignia of the order, and if not the distinctive coat of arms, is the distinguishing coat on the back); it is frequently mounted with as much pride and consequence as a pair of epaulets. The application of the term has frequently led to laughable *contretemps*. In a recent case, the son of a titled and distinguished military officer received an appointment as clergyman in a parish surrounded by mines. On the day following his first visit, in writing to his wife, a dame of high descent, he said—"I arrived safely last night at this secluded spot: although it appears to be a bleak barren waste, to my great surprise I find we are likely to enjoy most excellent society, and to be exceedingly gay; it seems positively to be a complete colony of naval officers, with numerous families, who, I suppose, have settled here for economy sake, as I assure you I have as yet seen no residences to bespeak anything more extensive, though there are no less than seven captains and five pursers resident within two miles of us." Poor man! little did he suppose what he in his idea had been describing was so nearly literally true in some respects— that they had large families, and practised economy! Little did he suppose the captains and pursers rejoiced in the splendid full pay on active duty of 6*l*. or 8*l*. per month; he little supposed they were poor, unsophisticated humble miners, who knew as much of the refinements of society as he did of their underground qualifications or arrangements.

To fill this important office properly, the individual should be possessed of certain natural physical capabilities, one of the most important of which is a robust, sound constitution, as exposure to wet, and the fatigue of climbing the various pitches, backs, and levels, where the miners work, entail considerable bodily exertion. He should possess a quick eye, a retentive memory, and keen power of observation; these endowments should be improved and cultivated by a careful and suitable education, embracing more particularly the liberal sciences of mathematics, chemistry, geology, mineralogy, civil engineering, and a knowledge of accounts; in addition to all these, he should have extensive practical experience, so as to adapt them to his purposes. It may be doubted if persons having such attainments are to be found filling such situations; we assure our readers they are, but certainly not so frequently as desired; yet among this class of professionals an amount of intelligence and ability in these branches is to be found that astonishes the stranger.

The duties of the captain are so multifarious as not to be credited unless they were enumerated; they are seldom considered, or rather are frequently unknown, to many adventurers, when voting the amount of their salaries. A false economy is frequently exercised by limiting their stipends to the lowest scale, as the best men will get and deserve the best prices. First, then, he has the whole duty of the mine devolving on him; if anything goes wrong, anything that could by foresight or forewarning be prevented or avoided, except by sheer accident, the blame is charged to him, and he takes the responsibility. It is his task to see the men do their work properly, and secure the safety of the mine at the cheapest rate,—to value the prices at which the bargains to raise the minerals, or to develop the mines, are let to the miners, so as to prevent undue expense, and yet let the miners have average wages,—to select the best workmen in the various departments he can procure,—to see they keep their proper time on the mine; if necessary, teach the best and most improved methods of dressing and returning the ores,—where only one is employed, he has to measure all the ground the men excavate; in many instances to make out the cost-sheets and to pay the men, in all cases he has to see this done correctly,—to keep an exact account of all disbursements made in the mine, and to see the quality and quantity of the stores supplied are in accordance with the market prices, and sent in as per contract,—to see the samples are fairly taken, and the ores correctly weighed off. He must also dial the ground, keep up working plans and sections of the mine, and lay out the positions for the various shafts, winzes, cross-cuts, &c., so as to be in their proper places to connect level to level,—superintend the erection of machinery in or on the mine,—furnish weekly reports of the state of the works, and appearances of the veins, receive visitors and deputations—attend committee and mine meetings—advise what quantities of reserved discoveries should be wrought out, and sent to surface for sale,—in case of accident, to remain on the mine till all is secure, frequently keeping him 24 hours at a time in the mine—attend the ticketing dinner (the most agreeable job of the lot), answer ten thousand letters of enquiry by inquisitive adventurers respecting his own and fifty other mines, with many other little jobs an auctioneer would classify as too numerous to be inserted in our limits.

It will be seen, from what we have stated, these functionaries have ample need of the qualifications we before stated; if it be an extensive mine, the proper performance of the whole duty is simply impossible. As many as six or eight are sometimes employed on one mine, to each of whom separate departments are assigned, who are amenable to, and must be superintended by, the captain—they, also, are named captains (we have heard of first-lieutenants of a windmill, but never heard the title applied to a mine agent, though they have pursers and mates). The captain *de facto*, is entitled to the prefix of head, or chief; the others are usually called underground captains, captain of the dressers, &c. "The chief" is, as may be supposed, a personage of no small importance among the workmen, by whom he is generally treated with the most profound respect, yet with a certain degree of familiarity—*i.e.*, where the captain respects himself, and takes care to keep the men at a proper distance. It is highly gratifying to see the discipline, and at the same time friendly feeling usually displayed on a well-conducted mine; the men looking up to their captain as a friend and protector, he on them as friends who have need of his assistance and advice, which is always cheerfully given. Most of these gentlemen are of studious habits and exemplary character, devoting most of their spare time to reading, and their Sabbaths to teaching the mine children in the Sunday schools, or preaching to the miners in the chapel. Some of them, by the opportunities they of course have in making discoveries, become adventurers, amass considerable fortunes, and retire; the majority, however, are a humble and particularly contented class, who owe their advancement and position to their intelligence and industry. They, too, have their bitter mortifications and insults, by silly orders from committees of management and snobbish chairmen and secretaries, who frequently are as ignorant of mining as they are of economics, who, from their own narrow-mindedness, think everybody like themselves, incapable of doing a good action or a disinterested service, frequently displaying a jealousy as unworthy a body of gentlemen, as to the high-minded fellow in their employ. They, also, often have to finesse even to keep the mine from being destroyed by the cupidity of the proprietors; instances of both these kinds of annoyances have lately come under the observation of the author.

In the former case, one of the shareholders in a mine had the meanness to dress himself in a dirty (ay, every way dirty) disguise, and visit the mine as a distressed labourer seeking employment, even going so far as to try to get employment, though he knew nothing of the business, and slunk about the premises for days, watching every word and motion—endeavouring to entrap the unwary captain, if possible! It would have served the fellow right had they ducked him in the pump stream or engine pool, or had (as he went underground too) played him one of the miner's coarse practical jokes sometimes indulged in; had he been caught in such guise in some mines he would have been roughly handled, and would not have escaped scatheless. In another, they had for years been paying a bi-monthly dividend of 10*l*. per share on a total outlay of about 5*l*. As usual, the same amount was to be declared, but it being found there were sufficient assets to increase it to 12*l*. 10s., nothing would do but divide it there and then, to the great inconvenience of all the agents connected with the mine, as there was scarcely a sixpence in hand to meet current expenses.

So eager are adventurers generally, especially those who are unacquainted with mining works, that had they their own way, they would far more frequently ruin the mines than they now do.

The author, a short time ago, was shown a splendid course of rich lead ore in a mine by the captain, who observed—" You must be very moderate in your estimate of this end, for if our adventurers knew this was here, the mine would not work six months; every end would be stopped, and all this exhausted and sold by that time. I keep up my sales and dividends out of this spot; if I pick out her 'eyes' she will stop in a week; for, although I have given them 60,000*l*. in dividends, a call of sixpence a share would 'knack the bal'." It may be satisfactory to know the mine has since become very productive; the judgment and wisdom of the captain has been the addition of at least 100,000*l*. to this property.

It will now be asked—Are these pictures correct? We answer—Yes. Every one is drawn from nature—from veritable facts. It will be said, also—Where shall we find such excellent and model captains? We reply—Go to Dolcoath, Devon Consols, Basset, Wheal Vor, Botallack, Wheal Busy, United Mines, or any similarly extensive concerns, where salaries are high enough to encourage and enlist talent, and you may see them as here painted, *veluti in speculo!*

## THE PURSER

AFTER THE CAPTAIN, THIS GENTLEMAN TAKES THE NEXT official rank on the mine, and is, therefore, a person of some consequence in the count-house, which is, properly speaking, his department, in the office of which he reigns supreme. He is treated in all respects with the same deference by the men and *employés* as the captain, with whom he acts in direct concert, save and except underground operations. The situation is usually reckoned one of the greatest respectability, "esquire" being generally appended to the name of a mine purser or secretary; a regular emolument being derived, renders the post much sought after.

To those who are emulous of taking the office, we commend an attentive perusal of our photograph, "The First Dividend," as the pleasures of life under such circumstances are correctly pourtrayed; we mean in a mine in which protracted calls are necessary. *Quem Deus vult pendere prius dementat* should be well considered; the age and truthfulness of the axiom entitle it to respect. Ere they take the fearful step, let the bold adventurers think twice. If after such monition they will do so, we say, be ye therefore wise as serpents and harmless as doves; and even then we defy them to give satisfaction; to be all things unto all men is the smallest, easiest part of their duty. To be anything to irascible people under certain circumstances, particularly when demanding money rather urgently, requires a command of patience and temper possessed but by few, yet this is absolutely necessary to fill the much-coveted office.

Notwithstanding the duties of secretary and purser require a certain amount of knowledge of mining detail and practical experience thereof, as well as of the laws, customs, and usages of mines and miners, and the Cost-book System (if the concern be under that peculiar regulation), together with many other requirements, only to be obtained by long residence on or large experience in mines; yet no sooner is such a situation open, but it is applied for by persons of almost every grade in society and variety of profession, from the thoroughly practical, qualified man, who really knows the duties he proposes to undertake, down to the peripatetic attorney's or auctioneer's clerk, who would be as well qualified for and as readily undertake the office of Archbishop of Canterbury, or leader of the forces in the East, grounding their claims wholly on being able to write a good hand, and competency to keep books by double entry, knowing as little of the complicity, formula, or detail of the duties necessary in this as in either of the high dignities quoted; yet they vigorously urge, and pertinaciously assert their abilities to "soon get into" what they do not understand. That is all very well, but how are they to get into the good graces of their companions on the mine, or of the committees they serve (themselves requiring information), if they do not thoroughly know and can practise their duties? It is an office of considerable difficulty, requiring a certain natural adaptation to fill it satisfactorily, yet "fools rush in where angels fear to tread."

Sometimes, to get in the thin "edge of the wedge," salaries are accepted which appear simply ridiculous in the hope of their being gradually raised, until they amount to something like a living value. Sometimes the situations are conferred on equally improper persons, by means of family or individual interest amongst the adventurers, when, as may be supposed, the remuneration is ample, and any slight omissions or faults overlooked. In the majority, however, in these as in all public appointments, favouritism and patronage prevail over ability or honesty, though the proper fulfilment of these offices conduces much to the harmony of proceedings in, and the welfare of the mine.

The duties of secretary and purser of such mines as have a London, or non-local management, are now usually divided, the pursership being commonly vested in the captain, who is provided with a clerk's assistance, to pay the merchants' bills, miners, labourers, &c., and receive the tin bills, &c., forwarding them and all vouchers to the London secretary proper. In others, a resident "mine agent" performs these duties for several mines, at a small salary from each, a decidedly advantageous plan, as it prevents the captain's time being too much infringed on by making out cost-sheets, paying men, &c., his work, if properly attended, to being quite sufficient to occupy all his time. Were such plans more generally adopted, and the number of mines superintended by one captain limited, fewer complaints of carelessness of measurement and absence of regular reports would be made. It is perfectly preposterous to suppose any captain, however clever, can superintend half-a-dozen mines, some of them miles apart, be continually reporting on others, and do justice to all; yet such there are, to the manifest injury of mining generally. This subject is attracting the notice of many large mining capitalists, who make it a *sine qua non* that their captains and pursers undertake the management of no mines but such as those they are engaged in by them, nor allow them to mix themselves up by inspecting and reporting on other mines; they engage, and expect to have, their sole and undivided attention to their own mines, which is quite enough for one person, and they pay them accordingly. The same rule is being applied to secretaries of large mines, where also they are debarred from dealing in the shares, a very proper regulation.

The salaries for such purserships of small mines vary from 2*l.* 2s. to 3*l.* 3s. per month. Their task may be described as consisting of visiting the mine two or three times during the month, for the purpose of forwarding authentic statements to the principal office, to consult the captain if he require any aid (such persons are usually practical miners, and quite equal to their places), to keep up the accounts on the mine, make out cost-sheets, attend on pay and setting-days, and to make out a monthly balance-sheet to be forwarded with the vouchers and receipts, thus considerably relieving the captains of such mines from, to them, the most puzzling and time-taking duties.

The multifarious services required of the "secretary proper" in London, or on the mine if the management be local, are but little known or understood by aspirants until the Rubicon be passed; they are of no easy or simple character even in the best and most prosperous mines. He has first to provide a well furnished, comfortable office, in some good eligible business situation (in London, or in the town in which the management is situated), to be there himself almost continually, if not, some properly adapted person instead, to answer a volume of questions by hourly-enquiring anxious shareholders, by whom he is expected to be as perfectly acquainted with the details, hopes, expectations, and probabilities of the mine's success as the captain himself; to answer numerous letters from distant shareholders to the same effect; to keep a complete set of books appertaining to the business of the mine, where from 5000*l.* to 10,000*l.* a year are turned over in wages and material, and where the name of every recipient of sixpence has to be entered; in addition to which are the report-book, into which the reports from the mines are copied, a book in constant requisition, and the transfer-book, also daily—nay, hourly—required by inquisitive shareholders; then there is the acknowledgement of transfer, and notice to new shareholders, altogether involving an amount of pen work little appreciated or supposed; to issue notices for and attend all monthly, quarterly, general, or special general meetings, as well as their frequent adjournments; to be at the beck and call of the chairman and committee, and woe to the wight if the mine be making calls! He then has to apply, badger, tease, frequently to threaten, and prosecute some unfortunate adventurer, in doing which, even if successful, he makes him an enemy for life—though the fault be not his he is charged with it; if not he is sure to make enemies of the whole of the committee, who at once charge him roundly with want of energy or interest in the affairs of the company; it, therefore, behoves him to be firm but courteous. He, too, poor scapegoat! has to hear and bear the terrible biting sarcasms and reflections of disappointed shareholders, as well as the hasty, hot expressions of impetuous purchasers, if the prices of shares go down, instead of up, as they anticipated—remarks the captain is, by his absence, frequently spared. He, too, has to listen to the noisy, egotistic twaddle of every Tom Noddy who pretends to understand mining, whilst condemning the whole management from beginning to end; and because he has purchased a few shares in it, claims and noisily exercises his right to overhaul the books for years past, making more fuss and ado than all the other adventurers put together. Such a busybody Heautontimorumenos is to be found in almost every mine; and woe betide the secretary, purser, captain, chairman, or committee, as long as he is in it! Better is it to buy the knave out at once, or he will ruin the mine, and set all concerned in it by the ears. If the mine be prosperous all goes on smoothly enough; as long as dividends are payable the posts of secretary and purser, separately or united, are all that can be desired; the management then do not object to a trifle, or grumble if that personage be off the mine for a day's pleasure during the season; a very different version is put on all his acts. If any of the adventurers visit his residence, and partake of his hospitality, as far as it goes, he is in the one case a "splendid, liberal, right sort of fellow;" in the other, "something must be wrong, or he could not do as he does,—we'll curtail his salary." Facts!

The post in most cases is one of considerable responsibility; very large sums of money are continually passing through the hands of these officers; in some mines they are constituted the treasurers, the money being lodged at the company's bankers at the call of the secretary, by whom cheques for such sums as are required being drawn by order of the committee, in which cases security to a considerable sum is required. Instances of defalcation to serious amounts do certainly sometimes occur, but they are very rare, perhaps more so than in any branch of business where similar sums are involved. The greater part of mine banking accounts are vested in the committee, by two of whom all cheques must be signed, as well as by the secretary, who, by signing these, renders himself liable to all sums withdrawn.

In Cornwall, the purser usually has the sole management of the mine and moneys (save the captain's office), and is frequently in advance to the mines many hundreds at a time. This plan has its advantages: as the secretaries there are generally well known men of substance, the bankers have little hesitation, and the men are regularly paid; whereas, under the former plan, if there be no assets, the committees are very unwilling to become personally responsible to bankers, the mines are thus inconvenienced by non-payment of calls, much to their discredit and injury. Many most honourable names are to be found amongst gentlemen holding these offices, celebrated alike for their integrity, ability as miners, high standing in society, and for kindness and affability to the captain and men. Among such stand the names of Taylor, Davey, Beckwith, Richards and others, as beacons for example and imitation.

Such, then, are the duties and difficulties devolving upon all who would aspire to become secretaries and pursers: we now appeal to the reader if we have overdrawn the qualification, when we say it is necessary to be as wise as serpents and as harmless as doves.

And now committeemen, adventurers, consider well, when appointing your officer, what his duties are, what his qualifications and abilities should be, and in proportion grant him remuneration; be not false economists, but where you have a man emulous to become what the Taylors and the Daveys are, give him a salary to encourage him to do so, by which means you will not only raise your class of officers, but your mine itself, in the estimation of the public; as the ability and character of the secretary of a mine has a great deal more to do with its welfare than many persons suppose.

## THE DOCTOR

WHEN AN ARTIST, BE HE A PHOTOGRAPHER, PORTRAIT, OR landscape painter, undertakes a work, he usually places his subject

in the best possible light. If there be any blemishes that he can hide
by advantageous position he exercises that discretion, which liberty
custom allows, as long as the likeness be faithfully preserved. The
aim of all pictures, we presume, should be to please, provided
fidelity be not sacrificed; in caricature, however, the order is
reversed, the peculiarities or defects in person or character are
exaggerated and distorted to suit the taste for ridicule; still, unless
the original characteristics be faithfully adhered to, the labour is lost.

We thus premise our portrait of the "Doctor," as we intend
transferring him as he appears, in the same style we did the "Captain," as captain, and as we probably may do in the case of the "Bal
Seller," in his turn; at all events, we ourselves adopt the advice we
tendered the captain, and remember the oft-quoted fable. The
doctor, then, is one of the most important personages in the parish,
be the mine where it may. He is always addressed by that title;
whatever his standing in the medical world, he is at home, *par
excellence*, the doctor, whether he be a duly diplomaed M.D. or
only a M.R.C.S. His emoluments are certainly somewhat considerable for a country practitioner. If the mine be extensive, or if there
be many mines in the neighbourhood, a few mine doctorships is of
some moment; for, though attending poor persons, his pay for
services is sure, which is more than the profession can boast of in
many districts. Their being located, as in the instance of East Wheal
Rose, at a distance of at least eight or nine miles from the nearest
town, is a great benefit to the agriculturists in the neighbourhood,
saving them large expenses for their travelling, besides the advantage
of immediate assistance; they, therefore, usually have an extensive
and valuable practice from these sources, in addition to their mining
engagements. Besides being the doctor to the mine, the situation
implies, as a matter of course, an attendance on the families of the
agents and men, giving him an augmented and remunerative
sphere of action. The frequent occurrence of accident affords them
good opportunities for surgical operations, consequently a premium
of proportionate value is obtainable on taking a pupil. Gentlemen
of acknowledged skill and good family are frequently to be found
candidates for the office, which, however, like many others, is
oftener conferred by the power of interest than by the aid of merit,
though, on the whole, few complaints are ever heard by the employers or by the patients of dereliction of duty, either in kindness or
attention, on the part of their surgeons.

In endeavouring to obtain the office, it is almost a necessity that the aspirant should be one of the locality, or, what is far better, a stranger coming into it, and taking unto himself a wife from among some of the bonnie lasses of an influential family, with a little cash and large connections (always to be, and frequently is, done), when things are likely to go on smoothly enough. The Cornish people are proverbially hospitable and partial to strangers: the greater the distance the greater the charm, as was shrewdly remarked by a keen observer of mankind (the late Mr. B. Sampsons, of Tullimar), who, when advising young men to go courting, urged the recommendation to go from home; "the further off you go the more cash you may get; the converse may apply." Most of these worthies hold an interest in one or other of the mines they attend, and not unfrequently take part in a new project with this end in view, by which means they are almost sure to succeed. Being pretty well advised as to the state of the mine, they have good opportunities of realising at proper times. From such advantages combined, many of the older members of the profession are persons of considerable means and position, thus giving a strong impetus to exertion by the younger branches. In passing through most of the church towns, as the villages are usually called, three houses, of more than ordinary pretensions, may be observed; these to a certainty may be set down as belonging to the parson, the doctor, and the lord's steward, if there be not a lawyer—if there be, the rest who knows not?

The usual professional duties of the doctor, we presume, are too well and painfully understood to be referred to in this place; we shall, therefore, only say they have here frequently to be exercised in their most trying and harrassing forms, accidents of a harrowing description occasionally calling on him to exert all his sympathy as well as ability. He has to make frequent visits to the mine, though not sent for, and to enquire into the general health and well-being of the agents, miners, and children employed. It is not surprising that he (as he sometimes does) times his visits so as to be there a little before dinner on account, setting, and pay-days, or any occasion of a meeting on the mine, as we find it not only suits the doctor's tooth to be present at such, but it suits his interests, too, as he becomes acquainted and familiarised with the persons there, who may possibly aid him at another time. Oh! a good dinner and cheerful glass are often great stimulants to social intercourse; besides, as they sometimes jocosely remark—for even grave doctors joke at such times, indeed, they are always good humoured in the count-house—money is always handy.

The salary of the doctor depends on the extent of the mine—that is to say, on the number of persons employed. All the miners have the sum of 1s., and the boys and girls 6d. each per month deducted from their pay for doctor and club. The club is a fund allowed by the miners to accumulate in the hands of the adventurers, in case of sickness or accident during their engagement, when the mine allows out of this fund 1l. per month to the men, and 10s. to the boys and girls, whilst absent from their work, which sums are regularly sent to them on the monthly pay-day by the purser. The doctor on some mines also receives his pay monthly, but more frequently quarterly or annually; for this contribution he engages to visit his patients at their homes, to perform all operations, and provide the requisite medicines and attendance.

This arrangement of doctor and club has been found of infinite benefit to the working miner, as well as to the parish; to the former, the regularity, responsibility, and comparatively liberal payment from this source secures them a far superior class of medical practitioners, and a greater degree of attention than they could otherwise possibly command. These people have an intuitive horror and dread of the Union workhouse, and can scarcely be persuaded to enter an infirmary, except in extreme and protracted cases; their friends exert themselves to the utmost to eke out their earnings, by the assistance of the club money and a trifle from the parish, if it cannot be avoided, rather than go into the poor-house, which is considered the depth of degradation.

The post of mine doctor in some situations is truly no sinecure. In the long, dreary winter nights, over a wild moorland, with not even a solitary post or beacon of any kind, sometimes hardly a sheep track to indicate the way, not unfrequently in the midst of open, unprotected, yawning shafts (a disgrace alike to the miners and magistrates in Cornwall), the doctor has to ride or drive a journey of some six or eight miles, often buffetting the fierce biting wind or pitiless pelting storm,—(only they who have passed over one of these barren heaths can appreciate the task)—perchance to some wretched hovel, where he probably has to tie his horse to a stone, or stake, without shelter of any kind whatever, not a tree or vestige of a residence being within miles of the spot. After performing his duties, he has the dismal satisfaction of retracing his steps on his shivering, dripping, steed, himself in a very little better plight, yet we never hear him complain. Such cases as these are by no means uncommon. The mines are frequently situated at distances so remote

from towns and dwellings, that the miners themselves have to walk
eight or nine miles to and from their work; as they are employed but
eight hours out of the twenty-four they have ample time for repose.

Such are the remunerations and duties of these officers, many of
them, as we before stated, being men of high standing in society,
who display in almost every instance the utmost tenderness and
attention whilst ministering their necessary but dismal duties. As
may be supposed, and, indeed, as in gratitude they should, the doctors are amongst the most respected individuals in the localities.
To them are confided all petty complaints or difficulties; everybody
has a kind word for them, and they for everybody. If of long standing, they may be regarded as chronicles of their neighbourhoods;
to them are known the descent, marriages, connections, and circumstances of the whole district; in short, the history of all that transpires
is open to the doctor. After all, his is about the most independent
and most agreeable situation on the mine, seldom coming in contact
with the adventurers; and none of the chances of speculation
devolving on him, he bears not their bitter, though it must be admitted sometimes deserved, complaints, or has to trouble unwilling
contributors for their modicum of outlay, nor to encounter any of
the thousand and one petty annoyances the captains and agents are
continually subjected to; yet he has his troubles and difficulties as
well as pleasures, none of them, however, so discouraging and
disagreeable as devolve on the other officers. No wonder, then, he
should always be cheerful in the counting-house.

In these remote places, though these gentlemen on their ambling
nags, or humble dog carts, do not boast the magnificence of their
more distinguished brethren, who in our cities and large towns roll
about in their stately carriages and handsome broughams, yet they
are as universally and deeply respected and beloved by their patients,
and are treated with as much confidence and veneration as can
possibly be conceived, a fact it is our duty and pleasure to record,
and which they fully deserve, without applying in the most remote
degree the *couleur de rose*.

## THE LANDLORD

THE IMPORTANCE OF MINING AS A SOURCE OF NATIONAL
wealth is generally overlooked by persons who merely judge from

the returns as they appear when periodically published, and who do not take the trouble, or have not the opportunity, to search out diligently and enquire into all its ramifications; by doing which, however (and it is absolutely necessary to its true appreciation), at every turn something fresh occurs they had not foreseen or supposed. As a matter of course, it must naturally have been inferred we at some period or other should find it our duty to describe the parties to whom the first application is to be made ere the mine be commenced. Had we originally contemplated extending this series of papers to the number now announced, we should probably have placed this worthy's picture at the head of the list of persons officially connected, he being absolutely the mainspring of the whole work. In our description we must necessarily speak of him less personally, and more of his rights and duties than of his peculiarities, as the only connections he has in the latter are to be summed up in the general terms of trusting to his stewards, or an over-grasping desire to screw the poor miner out of his last shilling. There are, however, some rare examples of liberality, and it will be our duty and endeavour to show how mining prospers under the two systems. By so doing we hope the pictures, by contrast, may be effective; provided they be effective and truthful, the end will be served, and sufficiently artistic, even if laid on by a slap-dash painter, which, however, is not our practice, or indeed the practice of any painter producing works of sufficient merit in portraiture to be deemed artistic.

It should be recollected that before any dividend in mines be declared, or even before the mine becomes remunerative, that as soon as the mineral is produced a portion, termed dues (in Devon and Cornwall usually 1-15th part), is claimed by the lord of the manor on which the mine is situated; hence the term "landlord." These claims are of extremely ancient usage, and vary in an extraordinary degree in different localities. These, again, alter even in the same districts, according to the depth of the several mines. In Wales and the North of England the rates are as exorbitantly high as 1-8th, or even 1-7th of the gross produce. But it should also be considered that the mines in those mountains are generally near the surface, and wrought by levels driven into the hill sides, thus preventing the enormous outlay requisite in the former counties for erecting suitable pumping machinery for draining their prodigious depths. In some instances the Duchy of Cornwall, holding extensive property in mineral rights, has liberally consented to grant on as low terms as 1-40th part, to encourage the re-working of old mines, which have

been formerly wrought profitably, but abandoned through imperfect machinery or untoward events—thanks to the enlightened
endeavours of Warington W. Smyth, Esq. As may be expected,
these rates are of the utmost importance to success; and the characters for liberality or otherwise of the different land proprietors are
well known.

Some noblemen, who have their own and the miners' interest at
heart, on being applied to by the mine agents, grant them a licence
to work for a time, until they can form a company, to give his
property a fair trial; that grant having a restriction of dues to 1-15th,
or a sleeping rent if the mine be not producing ore, so as to enforce
something like activity. Such gentlemen, when satisfied the company
is formed for working the property, usually take an active interest in
the welfare of, and frequently some shares in, the speculation, are
content with moderate dues, and grant a lease for 21 years (the usual
term), without much trouble or annoying restrictive clauses. So many
schemes are brought out merely to sell the concern, and then
abandon all further prosecution of them, that the landlords are, in
a measure, compelled, in self-defence, to adopt rigorous clauses.
These noblemen are not above attending to their true interests: they
send for, and confer with these mine agents and their stewards as to
the best terms and probabilities of making the most judicious use of
their properties, by not exacting too stringent obligations, or by
granting too limited or too extensive setts (equally unwise proceedings). These are the parties to whom an experienced Cornish
company would apply to obtain a grant for working purposes.
From the habit of devoting a little attention to these matters, they at
length become quite *au fait* at mining, and fond of the pursuit. Then
they take up the science in good earnest, and see that their own and
the company's interests are both properly attended to, by sending
their tollers underground frequently, and having the working drawings and sections of the mines brought regularly before them, thus
knowing, and feeling as much interest, as they would if they were
walking on one of their demesnes, and conversing with one of their
sturdy yeomen on the state of his prospects. Here, as in all similar
cases, to whatever branch of industry we refer, "the master's eye
makes the horse fat." When adventurers find they have an active,
liberal landlord, who in a measure identifies himself with them, who
by example stimulates exertion and encourages confidence, take
vigour if misfortune befall them, and are spurred on to redouble
their endeavours: like a good general, the good landlord is always

in the van, and is generally the means of gaining a victory which else had been doubtful. These gentlemen attend the mine meetings either themselves or by deputy, hear complaints if there be any, check any tendency to error, by design or mishap, and prevent many disputes, which, like women's ire, waxes fierce by nursing. Even the very miners respect these men, knowing, as they do, they are their true friends. Others there are who adopt a wholly different course, trusting entirely to their servants, who, as certainly in these as in other cases, take care of themselves as well as serving their masters; hence the delays and the enormous extra expenses for their travelling, viewing, consulting, and agreeing (even then frequently a douceur or handsome premium is the only way to obtain a preference). Then the drawing of the lease is generally made so stringent as to hamper the miner. This over-straining is sometimes adopted to blind the foolish landlord, by the display of a studious anxiety to protect his property, when, at the very time, they are sapping its well-being. Again, there is often a desire to curtail the extent, so as to be certain of securing a second or third premium, and expensive leases for extensions and additions, together with several letters, consultations, &c., too numerous to mention, but which must all be paid for ere the lease be signed and delivered. The fable of the lark and her young ones is quite applicable to mining landlords!

If they do not attend they cannot be conversant with the miners' wants, or in what fair terms for them consist, nor do the shareholders take that genuine, heartfelt interest as when the landlord sets them such an example as the one we call liberal. It is really surprising what ignorance is sometimes displayed by such gentlemen when they do attend a mine meeting perchance, as in the instance of one Sir S. S——, who had been earnestly requested, on granting a new lease, to reduce the dues on a mine struggling against adverse circumstances from 1-15th to 1-20th. He was invited to attend their next meeting on the mine (which he promised to do), that he might have thus displayed to him the vast cost they had incurred, the time they had persevered, and the considerable sums they had already paid him for dues, though he had not risked a shilling, but they had lost their thousands of pounds. This honourable baronet accordingly attended, the adventurers met in full assembly, and forcibly made their appeal; they thought they were unsuccessful, for the worthy baronet enquired what Sir R. V—— received, who held a moiety of the mine. On being informed it was 1-20th, he made a speech (he had been in the habit of addressing electors), in which, after expressing his entire satisfaction of all he had heard and seen, he concluded by saying he owed a duty to himself as well as society, and that was to protect his successor as well as himself; that as Sir R. V—— was receiving 1-20th, he could see no reason why he should receive less; therefore, he regretted to state he should insist, before any new lease be granted, that his emolument should be advanced to the same terms as Sir R.'s, and that he should direct his solicitor to act accordingly! An announcement which, of course, was received with a subdued smile, and was cheerfully but silently acceded to; the same dues as Sir R. V.'s being actually now paid. Had this gentleman devoted a little experience to his duties toward society and successors, he would not have fallen into so glaring an error, or have made himself the joke and laughing-stock of the county!

It may be doubted whether in some instances it be worth the while of noblemen or gentlemen to bother or trouble themselves about such affairs, except they have mining properties which are working,—we say, it is so under any circumstance, as has been proved in a multitude of cases. A most remarkable one occurred in Wales, where a nobleman, who was in straitened circumstances, and considered a poor lord, is now in the recept of 5000*l.* a year from his mining property. At St. Ives Consols an eccentric old lady at her death bequeathed a paddock to her donkey for its support, at its death the property to belong to a faithful domestic. From this source the family have for many years been receiving large sums in the shape of dues, enabling them to become shipowners, and persons of considerable property. At Levant Mine, a gentleman who was only a curate, on very slender resources, has long been in the receipt of dues that would have founded a bishopric. At Tresavean, the Rev. Canon Rogers, and the other lords, have received 60,000*l.* The Devon Consols have paid the Duke of Bedford 100,000*l.*, and are likely to pay as much more, besides 20,000*l.* for a renewal of their leases.

These enormous revenues are derivable more or less from all mines; in some noble instances, however, the landlords who have their own and the miners and society's welfare in view, generously forego all claims until the shareholders have received some remuneration for their anxiety and outlay. Their lands are, therefore, much in request; but where it is known such is not the case, or where it is felt there is an exacting, grinding, half-lawyer half-farmer steward, who has too much greediness to be honest, and too much egotism to possess ability, the lands are shunned, or taken by speculators for gambling purposes, at once the bane and disgrace of mining.

The noble sums returned as dues enable many landlords to improve their estates and cultivate the wastes; the roads necessary to be constructed for the conveyance of ores and materials are frequently made at a cost that would seriously embarrass agriculture; by cutting the long adit levels, so generally requisite, a system of drainage is effected of incalculable benefit, yet these advantages are frequently overlooked, or not sufficiently appreciated.

Though foreign to our subject, as photographers of Cornish characters only, we shall be excused drawing attention to the conduct of the Marchioness of Londonderry towards her miners, and then think it not beneath our dignity to notice and cultivate these men or their profession. Therefore, oh, landlords! gentlemen of England, consider well the two sketches of the picture; we have little hesitation as to the result, and commend you by all means, for your own and your country's good, to "go and do likewise."

## TRURO FAIR

SHORN OF THEIR PRIMITIVE CONSEQUENCE AND UTILITY AS are the country fairs in all parts of the kingdom, by the innovations of modern taste, the improvement in public morals, as well as by the rapid increase of towns and villages, with their handsome shops, filled to repletion with goods suited for every requirement, few of these ancient substitutes for such accommodations retain a magnitude and celebrity equal to the coppercrist (*corpus christi*) at Penzance, and that forming the subject of our paper, held on Whit Monday. It is not our intention to write a chapter on the origin, rise, and fall of fairs, or to enter on Ossianic laments for their departed glories, great as they were in our childhood's imaginations, when the dazzle of tinsel kings and bespangled queens, with all their mighty hosts of attendants, in brilliant glazed calico robes, resplendent with pasteboard and glass jewels, surmounted by graceful, but dirty, nodding plumes, mounted on their piebald, highly-trained, but lowly-fed palfreys, with the nimble pantaloon and the dear old clown, cracking his still older jokes, afforded us infinite gratification and delight. Happy days, happy days! when mere appearances so pleased us. Yet, if we moralise or reflect, are we not still amused and taken by tinsel and appearances? Are we so very much altered by age? We fear not.

Even if we be, boys are and will be boys still; then let them enjoy their heyday, as we have enjoyed ours. The same *animus* pervades childhood as ever. At all events, "the camels are coming, the camels are coming! hurrah, boys, hurrah!" is still the burden of the cry of the joyous throng of all the children in and about the locality, as the cavalcade of performers, male and female, make their formal state entrance into the town, in their dragon, pagoda, Chinese, or some other outlandish, barbariously designed description of vehicle, with twenty horses, driven by the world-renowned, never to be equalled Mr. Emidy; Ramo Samee the second, or young Powell, perhaps a descendant of the celebrated troopmaster of our boyhood, preceded by the unrivalled band, with noses as red as the second-hand beefeater's garb in which they are arrayed, and followed by the magnificent stud exhibited in detail.

For a mile or two on the high road crowds of urchins, and children of a larger growth, amid heat and dust, wander out to meet the procession, each youngster fixing his fancy on a favourite horse, in their little hearts literally envying the pale faced, sottish-looking rider his glorious, gaudy, party-coloured trappings and crest—poor children!—each deeming himself but too proud to be distinguished as the bridle-holder of one or the other of the prancing, spurred-up steeds; the preference is given by all, of course, to the most splendidly clad, confident looking personage. Here let us ponder again, and consider that we differ now but in age.

Then come the group of strolling players, a squalid crew, with all their concomitants, beside the entire scenery, machinery, and decorations for a theatre; there are the wives, the daughters, and sons, the phenomena and Roscii of some future stage, provided the magistrates and spirituous liquors spare their existence; but there they come, ragged, dirty, redolent of hair oil, tobacco smoke, and gin, presenting as motley a group as in the days of Hogarth. After these come the lions, tigers, and animals of all possible and impossible descriptions, the most savage being their keepers and showmen, with their hoarse, husky voices, and filthy, rank smelling, blackguard assistants and drivers, who keep up a running comment of varied swearing and slang, picked up by long practice and experience in the most notorious sinks of iniquity accessible during their peregrinations. After these, on every road, may be seen the small fry of the profession, including conjurors, giants, dwarfs, learned pigs, and photographic professors; *theatres du petet lazara* and lazier proprietors, swings, up and downs, thimble riggers, blind or pretended blind fiddlers, beggars, and a detachment of Falmouth tag rag and bob-tail, three or four score. This influx continues, with little intermission, from midnight until midday, during which period all hands that can be spared from the attracting procession are busily employed in fighting for and fitting up the most eligible situations on the "High Cross." Having thus, as we presume, properly premised the day, we shall enter on our duties with greater ease, and be enabled to depict the scene with greater lucidity. The dusty roads at early morn are dotted by flocks of sheep and herds of cattle, with their attendants, of far more useful though less pretentious character than those lately passing in review; these wend their way to the Castle Hill, if not purchased on their route by forestallers; if so, these poor over-driven beasts, have to stand for hours broiling in the sun, and harassed by continual blows from their keepers, until their chiffering (haggling) proprietors adjourn to a neighbouring pot-house, to settle their differences over a pint of toddy. The cattle fair is over by 1 o'clock P.M., after which the country maidens (all lasses are maidens in Cornwall), muster in full force at the High Cross— *spectatum veniunt venicent spectentur ut ipsi*. As a natural consequence, the young men are to be found in proportionate numbers; a fine sight, indeed, they afford. Few counties can boast a finer display of honest lads and bonnie lasses, with their smiling, rosy cheeks, and modest demeanour, and brawny, stalwart, handsome youths, than are to be seen at the High Cross on Truro fair day. These form the real *couleur de rose* of Cornwall, without the use of paint; it is natural. We know we shall have the opinion of those who have seen them in our favour, despite critics.

Meantime, the older yeomen, captains and agents of mines, merchants, &c., repair to the different hotel ordinaries, to talk over politics, prices, old acquaintances, and minor matters, and to enjoy themselves, this being a day of general recognition and mutual good humour; the third, fourth, and sometimes—but we and they forget how many glasses extra are indulged in; no matter to them, or to us, so that their horses know the road home. At 2 o'clock P.M. the pleasure (?) fair commences; the *debut* of the *corps dramatique* on their narrow stage is anxiously awaited by the throng; at length they issue forth, in all the panoply of dingy, threadbare, cast-off wardrobe from some metropolitan suburban theatre, as incongruous a medley as are to be seen at one of Jullien's *bal masques*, though not so numerous. The hero of the Bloody Hand or Tale of Horror, the Demon of the Black Forest or the Remorse of Guilt, or some other horrible affair (the more horrible the title the better the effect), announce in stentorian voice that they will on that occasion present the astounding novelty of one of the above thrilling tragedies, with an interlude of singing and dancing, the whole to conclude with an entirely new pantomime, entitled Lodge Secrets, or Harlequin Jackanapes, and the Freemasons' Gridiron, showing how easily a fool is made, and all for the sum of one penny. Only think, for one penny! a real tragedy, a song, a dance, and pantomime, and only a penny, the whole executed too, in 25 minutes! Could Mr. Kean be keen enough to beat this, even with an episode of fools?

After an emphatic eulogy on his brother actors, and modest allusion to his own abilities, by way of commencing business he considerately horse-whips Mr. Clown, who was doing his duty by caricaturing his words and grimaces, which chastisement the latter duly avenges upon Pilgarlick, to the infinite glee of the delighted crowd. After many assurances that now is the time to secure good places, a few stragglers, like decoy ducks, gradually enter; some voice in the crowd bawls "give us a dance; let us see what you can do;" when Mr. Merryman's services are again called into requisition he being M.C. of the *al fresco* drawing-room. Then comes forward a queen, but whether of Henry or Hector the Second can scarcely be distinguished by her garb; to judge from her pallid, wan countenance she must belong to the latter, for, poor creature! she seems to have suffered hectoring enough from some tyrant! Then there is Columbine, not more than sixteen, in pink "tights" and starched muslin—*qui color albus erat, nunc est contrarius ablo*—reaching nearly to her knees; poor child! her lot is indeed to be pitied! Then there are Circe, Venus, and many other goddesses of similar character, ranged opposite the Duke Aranza, Hamlet, the Stranger (alike to soap and sentiment), with many other walking gentlemen; the most perfect characters being the villain and Jeremy Diddler, both up to the mark in real as well as mimic life. These join in a dance unknown to any but such professors of the Terpsichorean art. After a display of considerable agility on the part of Harlequin and his partner, the music stops; the hero, with stentorian lungs, orders "All in—all in and begin!" again admonishes the crowd,—now is the time to secure places as the curtain will rise in five minutes, on such a spectacle as was never before presented on the histrionic stage (being probably true).

The audience, now waxing warm with excitement and heat, throng in in numbers, the place is soon filled. As it is fair time, and only a penny! suppose we go in and witness a performance? Well, there is the drop-scene, once well painted, and, if we recollect aright, we once saw it at the Grecian Saloon, but how sadly worn and disfigured. We secure a seat in the boxes. What's in a name? boxes and pit! Oh! ye gallery gods! spare us a little of your jocund, boisterous hilarity and gratuitous showers of orange-peel whilst we endure this horrid suspense, and still more dreadful din called music, produced by two violins, a clarionet, a drum, and pandean pipes, the latter worked by a one-eyed black fellow with provoking energy. Complaints are useless; it is fair time, and all jokes pass in fun, fun being the order of the day.

At length the curtain rises on a scene literally as had been promised; there was the hero, surrounded by a most mysterious-looking forest and rocks, who, in "language brief," described the secrets of his prison-house, whence he had just escaped (probably, partially correct), with the most approved nasal utterance and stagetic tread. The heroine with the wan countenance (whom by her dress we had mistaken for a queen of some sort), appeared with an equally short history of her life, trials, and present misfortunes, when the villain makes his entrance, and seizes the lady without any ceremony, upon which the hero rushes furiously upon his victim; a terrific combat ensues, both fall; the demon, amid the full radiance of a pot of blue and crimson fire, appears, and walks off with the lot,—an emblem of the consequences of guilt they would do well to recollect. The curtain now falls, amid such shouts of applause as Mr. Punch receives when he leaves his place in Fleet-street by his wooden representative for a "tour of the provinces."

Before the curtain is well up, the comic singer, frequently in a miner's dress, hops on the stage and performs his part. A great favourite is an old song, of which we retain one verse, running:—

> "If you wed a miner,
> He to bal will go;
> He'll come in on a fair day
> And take you to the show."

This exquisite *morceau* of lyric poetry being executed with suitable grimace, is certain to call down an encore, which time will barely afford; still, it must be complied with, the public enforcing their rights to double the promised quota here as well as in more polite assemblies. The dance, a *pas seul*, also secures an encore; when the "Lodge Secrets" commences, and elicits roars of laughter, heard outside the walls (canvas walls), for by this time the hero is on the outside stage shouting, "Hear them! hear them! Only hear their approval of the performances!" which appeal and confirmation secures a rapid refilling by the visitors, now all anxious to get good places, of which they have as much chance as herrings in filling a barrel.

Genii of the stage! shade of Grimaldi! or thou, O Momus! grant us but power to describe the pantomime. We think we hear you say, "We are not Freemasons, go to them." Therefore, we appeal to you, oh! ye brotherhood! But no, we should be letting the cat out of the bag, and spoil the poor players and masonic farce at the same time. We will not do it; let all pay for knowing; experience bought is better than experience taught. Suffice it to say, the whole assemblage laughed till their sides ached, and tears ran out of their eyes; we confess to the soft impeachment, we could not resist, so truly excellent was the burlesque. The whole performance extended to full five-and-twenty minutes, long enough, in truth; the comic singer returned thanks in brief terms, assuring us of the manager's thanks. We were soon out of the house, and in five minutes after we heard the bell tinkle for commencing *de novo:* on looking at our watch we found we had been detained just thirty-five minutes by our curiosity. To see all its phases, we took a stroll round the fair, and found the whole of the professionals in full activity; the noise, clamour, and rivalry, worthy a Greenwich or a Bartholomew exhibition in their palmy days. We visited the resorts of the labouring class and the miners; these we found in the public-houses, enjoying themselves with their wives and sweethearts to their hearts content; rum, shrub, and ginger-beer, being substituted for the once favourite but now almost obsolete "gin and treacle", and hot "beer and sugar made sweet and fulsome" (formerly the miners' *beau ideal* of liqueurs). Some parties we found singing songs, but more of them singing hymns,—an extraordinary taste certainly, but literally true. As a matter of course, some, like their superiors on such occasions, got "fou", and as when the liquor is in the wit is out, a jolly row or two took place, a fair stand up fight, an interference, a reconciliation and shake by the hand ensued. These were the worst—indeed, the only bad features in this celebrated miners' fair; a contrast, indeed, from thirty years since.

There, too, were the mountebank doctors and vendors of pills, immortalised by Richards in his *Cornish Dialogues*, where they are described as—

"One for curing sore lips and sore eyes,

And one for giving (?) all sorts of disease.
It cured Jenny's leg when 'twas rankled and swelled,
And the back of the moyle when 'twas terribly galled."

These once famous remedies are, however, fast falling into disuse.

The revelry is continued until about eleven o'clock, after which the leading players and showpeople retire, "God save the Queen" preceding the last performance, the audience consisting of only a few townspeople, the country population going on their route at a much earlier hour; as the clock strikes twelve the people's day is over for another year.

We have portrayed the good and bad of this rustic festival as they exist: we know many strict moralists urge the entire suppression of such affairs; for our part, where they are confined to such hours and scenes as we have depicted, we should be sorry to see them discontinued. "All work and no play makes Jack a dull boy;" and if a few overstep the bounds of discretion, they form the exception, not the rule; the innocent enjoyments of the whole for once a year should not be blamed or jeopardised. We confess the experience of the day, with its lithe jollity of the youthful, the hearty congratulations of the elders, and the delighted exuberant glee of the children, determined us to recommend our friends, and to ourselves revisit, as often as opportunities offered, the gaieties, frivolities, and amenities, of Truro Fair.

## THE MINER'S WEDDING

IMPROVIDENT AS IT MUST BE ADMITTED SOME OF THE working miners are, it must not by any means be supposed the generality of them are so, many among them displaying anything but that coarseness of conduct their rough exterior and laborious employment would lead a stranger to expect, presenting a striking contrast to the "navvies," whose daily vocations in some measure resemble theirs.

To do full justice to our subject it will be necessary to present two photographs of the same, taken from different aspects; the one will be found in the taste and in the manner of Teniers and Brauwer, the other more resembling the pictures of Watteau or Andrews. Though so perfectly dissimilar in style, they are, nevertheless, correct, the period of time in some degree causing the difference.

To those who know what miners' residences were in Cornwall 30 or 40 years since, to the traveller who has noticed the wretched straw or rush thatched hovels frequently passed in the wild moorland districts, or to those who have seen the impromptu erections of the navvies on railways, or to Irish and Welsh mountain excursionists, these cabins need no description; to those who have not it may be necessary to state that they usually consisted of a room, sometimes two, formed of walls generally built without mortar, and of the very rudest masonry; not unfrequently by the miners themselves; or by a hedger, in the same manner as he erects the hedges dividing fields. In these there were seldom any kind of flooring but the earth, and no ceiling but the thatch, supported by rafters of the sawn Scotch fir poles or old mine timber, and secured thereto by straw spun ropes, from which large stones are hung to the eaves. In each room was a window of four panes of glass, about 10 inches by 8; seldom were they larger, glass being then very expensive; frequently an old black bottle built into the wall with the neck outward was substituted. A chimney and ingle (or chimney corner as it is called) was always provided, as was a horse-shoe nailed over the entrance, to protect the house from "piskies" (fairies). There, too, as certainly were to be seen the piggery and horrible mudpool in front of, and in immediate proximity to, the door. Sometimes these wretched hovels were built of "clobb"—i.e., clay and straw. These were reckoned of superior order. If of two stories, were usually walled stone up to the first and the upper part only of "clobb." These were of comfortable dimensions, and were really pleasant cottages, enduring for many years. Vast numbers of houses are even now built of these materials in Devon and Cornwall, though gradually giving way to stone, brick, and slate. In such huts as we have before described (a few scattered here and there may still be seen) have many an honest couple brought up numerous families without repining, or even deeming theirs a hard lot! To such places have many a youthful bride been taken, deeming herself but too proud and happy to be its mistress! The custom for colonies of families to live in a large house, as in many of our towns and cities, never prevailed in Cornwall, every one endeavouring, if possible, to have their own dwelling. As soon as a young man is able to support himself he was expected, and it is not surprising he was glad, to seek a fresh abode, either at lodgings, or marrying and keeping house. These circumstances led to the habit of early marriages so prevalent in this county, where youths not out of their teens were frequently parents. A miner lad's pride and care is first a watch, then a gun, a clock, a chest of drawers, a bed, and then last and best a wife; of these almost every miner in the two counties is possessed, besides olive branches. The wedding-day was generally postponed as long as decency would permit; this was the rule not the exception, and was carried out with unblushing effrontery. Though such important and rapid improvements in the morals and tone of society in the lower orders have undoubtedly taken place within a few years, this shocking laxity amongst them is still to be regretted. It is, however, much less the case now than formerly. Fewer instances of illegitimacy occur in Cornwall than any other county, liaisons generally ending in marriage. We should not have alluded to this subject, but truth must be told, and may perhaps do good by exposure. At the time referred to the tastes of the people were exceedingly gross. The wedding was then little better than a drunken spree for the bridegroom and his companions, continuing their orgies for three or four days, or until their little savings were expended. After the ceremony (properly so termed) the wedding party repaired to the public house at the church-town in their best array, where a fiddler or two were provided; eating, drinking, dancing, and smoking being the order of the day as long as they could stand their effects. Drunkenness of course in the youths was the consequence; and many an acquaintance formed on such occasions led to the necessity of a repetition of the scene a few months subsequently by more than one couple of the "young people." No better opportunities of amusement then offered; no one sought to reprove their evils or improve their morals, so on they went, drinking and rioting until their cash was out, and then to work for more.

This is our Teniers or Brauwer picture,—now for the Watteau-like representation.

Unless persons have actually witnessed they can scarcely believe the amazing change that has come over these people in their habits of domestic comfort, personal appearance, and demeanour. The females are now particularly neat and tidy; on Sundays and holidays they dress remarkably well; indeed it is surprising to see their gay attire; true they have rather a reprehensible taste for gaiety and display, but not of that gaudy, vulgar, description some localities present. The men, too, are well dressed; usually in good broadcloth suits; of course we now speak of the steady and more respectable portion of the community, who are so numerous as to form the rule.

The temperance movement has been one of the main instruments in this mighty change. In almost every mining town and village are houses and terraces built by these people out of their hoarded earnings from this most admirable source. Thousands, by abandoning the glass, have been enable to secure the means of emigration, and after a few years in Australia or California to return with a competence, or remain in the colonies as wealthy settlers; their mining experience being there turned to good account. Many mines in Cuba and South America are managed by Cornish agents and worked by Cornish miners. These return with considerable sums, as well as their tastes improved by travel and mixing in better society. All these circumstances, combined with the steady demand and regular remuneration for labour experienced of late years, together with the improved discipline of the church, chapel, and Sunday-schools, have led to and effected the amazing change.

The cottage gardening societies form another important element of improvement, and are rapidly extending. The tendency of such establishments is so obvious as to need no comment. A taste for flowers and the comfort of a kitchen garden have their due effect.

The squalid hut, with its mud floor, has given way to neat cottages with paved or brick floors, and "planch wooden chamber." Now there are always three or four rooms in their cottages, so that the families are divided, and that mixing of the sexes and huddling of children does not exist. Improved dwellings have much more to do with the elevation of character than most economists give credit for. The furniture is generally in consonance with the pretensions of the dwelling, as is also the cleanliness thereof. In this, as in all the rest of the domestic comforts, a marked difference can be observed, though much still remains to be accomplished; the intolerable cesspool being still an eyesore, and an olfactory offence; yet the progress is steady, satisfactory, and continuous.

The introduction of railways and picnic excursions has done much good. These are now the usual modes for spending the wedding-day; on any of the great festivals, such as Easter, Midsummer, or parish feasts; many a small party of six or eight may be seen off on such a trip in these conveyances. The Cornish Mount is one favourite resort. The Logar Rock, Land's End, or some of the beautiful coast scenes, are largely patronised on such occasions. As they are worth seeing, we will accompany a wedding party to one of them. From Redruth or Camborne they come by rail, as smart and hilarious as can be, with well filled baskets of eatables, and a fiddle or flute; many of the miners being good musicians, one is generally of the party. These wend their way to Marazion, where, as we will not arrogate too much, and claim all as teetotallers, they wet their whistles with bottled porter, rum shrub, or port negus; then off to St. Michael's Mount. After a ramble round the rocks, and an examination of the copper and tin lodes, there denuded by the sea, they visit the castle, and dare the dangerous, silly feat of sitting in St. Michael's chair[7]. After this foolish feat, amid much fun and bandinage as a matter of course if any have sense enough to refuse the foolhardy attempts, the party scramble down the narrow winding staircase, where again considerable merriment is created by groping their way in the dark passage. The chapel and dungeon being duly examined, the terrace visited for the sake of the magnificent view, and having the various mines pointed out (upwards of 80 being visible), the party repair to the back of the island, where, in some secluded spot, they spread their picnic meal, and enjoy themselves with jokes, singing, and music; after which they stroll about the delightful spot, *chacun a son gout*, until time warns them of the approaching train, when they hurry to the pier, and after a brief passage per boat to Marazion, they re-enter the hostelry of the "Commercial," where, by this time, tea, with all the necessary appendages, are provided. Another drain of grog and negus completes their extravagance. They then repair to the railway, and in an hour or so are at the "old people's," where a supper is usually provided, and the glass handed merrily round until 10 or 11 o'clock, at which period the young couple repair to their new abode. The bridesmaids and guests are seen home by their "young men;" each with a bit of wedding cake, to dream of their own wedding-day. Thus is spent that of the modern, well-conducted, Cornish miners, whose example we hope and trust will, as it is now, be daily and increasingly copied.

Our picture is now complete; and, though it be but a faint copy of the master we have alluded to, it is only defective in colouring (pardonable in photography). The refined taste is there, as well as the characteristic difference between rough, vulgar boors and the amenities of more educated and polished society.

Now, reader, this mighty change is true; and, as it has arisen solely from the causes described in these papers, it becomes your duty, as far as in you lies, to encourage and promote improvement in the dwellings of the working classes, to excite self-respect by improved domestic comfort and dress, and by keeping the sabbath—

that really day of rest in Cornwall—by aid of the church and chapel; so shall you witness not only greater improvements but the gradual extinction of the evils we deplore.

We will now wish our young couple goodbye, and all the happiness they wish themselves.


## ST. JUST FEAST

"IT IS A FAINT HEART THAT NEVER REJOICES," IS THE CHEER-ing apothegm to many a saddened, depressed spirit, when re-suming its wonted elasticity, after severe affliction, inducing that return of self-confidence so desirable and laudable in adversity. This exciting phrase has kindled many an ardent spirit into vigorous exertion, after extreme despondency. We do not, however, now intend to apply it in these its highest attributes, but as the watch-word and palliative for the exuberant glee, riotous festivity, and (for them) lavish expenditure indulged in on the ancient annual festival forming the subject of our paper.

Owing to the great number of mines at work in its immediate proximity, many of which are remarkably productive, this once small, and then secluded, village has so increased during the last 20 years that it rivals many of the older country towns in population and appearance, the old and handsome church and high cross forming prominent objects; its very large and stately Wesleyan chapel surprises a stranger by its proportions, exciting a wonder where a population can be found to fill such a tabernacle, in ad-dition to those passed in every hamlet on the wayside.

St. Just is the only village of any pretensions west of Penzance; it is, par excellence, the mining parish of the West, being full of mines, and conferring, for this reason, a name to the whole district; the people are generally miners; the land, from its proximity to the sea, and consisting principally of hilly, wild, rocky, bleak, common and moorlands, affords but small employment for agricultural labourers. Not a tree is to be seen; we believe that there are not a dozen in the parish: there were a few at Nancherrow, but we think even those are gone. If agriculturists be in the parish, they soon become imbued with miners' independent habits and associations, and go underground, if the miners will allow them. A regular miner plumes himself as being far superior to a labourer—as indeed he ought, seeing he executes as much work as two or three of them. The wages paid for mine labour in St. Just, until lately, were lower than in any other part of Cornwall, and the men not so well treated; emigration and oppression have done their work, and remedied this. Now St. Just chickens, as they are called, are to be found everywhere; and a hardy race they are—born on a wild common, inured to hardship and toil from an early age (at 10 years old boys and girls are sent to mine), and when they arrive at manhood are remarkable for activity and strength. The period of a miner's labour underground is eight hours; he, therefore, has several hours in which he can employ himself beneficially, especially in summer time, their cottages usually having a garden, meadow, or potatoe field adjoining, in which they amuse themselves; in the autumn, they form parties of two or three, and go out night fishing off Pendeen and St. Just Coves, where a sail or row of half an hour places them on some of the finest fishing ground in Europe. At this season, when the pilchards are off the coast in myriads, large fish abound also, attracted by the vast shoals of their favourite prey. Not unfrequently, two or three nights' fishing (these miners being very expert in the use of boats and fishing gear) suffice to supply their families with a winter's stock of prodigiously fine cod, ling, hake, and conger eel, which, when carefully salted and dried, are to be seen in every house, forming many a substantial meal; these, together with salted mackerel and the favourite pilchard, form the staple food of the inhabitants of this and other mining parishes.

Until within the last ten years, Cornwall was celebrated for the excellence of its potatoes; these were also abundant and cheap. Then, "pilchers and tates" one day, and "fish and tates" the next, with "flesh" on Sundays, formed the bill of fare throughout the year, except at Christmas, when "goose", and on feast-day, when roast beef and plum pudding grace the board, with a drop of "moonshine" [8] to wash it down, for which gratification many a brave and daring fellow's life has been periled and sacrificed on that dangerous coast! Could but the adit levels along the cliffs at Wheal Owles, Botallack, Wheal Cock, or Levant speak (they roar with the gale and whisper in the zephyr breeze)—could they and Pendeen Cove reveal the secrets transacted in them—Custom-house and Revenue officers would stand aghast, or perhaps would, as it is said they sometimes do—but we won't say what; that is, or should be, their look out, not ours. Everybody knows there must be "moon-shine" at St. Just—aye, and Buryan, too—for feast. Even if a certain old gentleman stood at the door and ran the risk of being put into a Cornish pie[9], "moonshine" must be had for St. Just Feast; it always was so, and always will be so to the end of time. We have tasted it, and will taste it again if we have the chance, were it only to keep the old custom up. It is the practice from Trewellard to Chapel Carn Bray, and from Cape Cornwall to Newbridge; it has been so from the days of St. Just himself, who, doubtless, like the Coast Guard, winked at its introduction, and smacked their lips at its taste, each being wise enough in their generation to "ask no questions." Had Father Matthew himself been there on feast-day, we verily believe he would have taken the *cacoethes bibendi*, and have "sworn there was nothing like grog." It is necessary to give this long prelude, that the lights and shadows of our picture may be more plainly developed. The habits of these poor people being, of necessity, very economical, their lavish expenditure renders the contrast more striking, and the zest with which they enjoy their festival the more keen. To many, who fare luxuriously every day, it may appear trifling and silly; to them it is a high day—indeed, the highest in the calendar, all parish gossip taking its date from its golden letter. Time, the revealer of all secrets, hides the feast-day until it comes; but it requires no almanack to foretell its advent; six months before hand children, unable otherwise clearly to define the period, begin on their tiny fingers to count up the weeks: the bal boys, too, begin to demand sixpence a month from "mother," out of their wages, for feast. The frugal housewife, too, puts by all she can spare for "nackins;"[10] and the men, too, work hard and long, to earn a little extra wages over their regular pay, so as to have a shilling or two to spend over and above the "old woman's" allowance (as "it won't do to let the miller know of all the water that goes over the wheel"), each worthy endeavouring to keep their savings as private as possible from the other, for fear anything unforeseen should turn up to prevent a feast. It is a time of rejoicing among friends and acquaintances—such a feast as we presume was intended of old, when none, not even the wayfarer or stranger within the gates was to be unwelcome—distinctions are levelled, animosities forgotten, and compliments exchanged throughout the length and breadth of St. Just parish.

Soon after noon on Saturday, crowds of the miners and their wives, in holiday attire, may be seen hurrying off to church town to make purchases for the important occasion; the latter, after making their markets, just take a little "drop" by way of commencement, and are soon off to prepare for the morrow. The church town is the scene of great bustle. The butchers provide the finest oxen they can purchase for this market, which is plentifully supplied with every necessary. The place is filled with "Cheap Johns," and all the minor appanages of "Truro Fair," showing all the outward and visible signs of a repetition of that spectacle. The public houses are filled— we meet half a score mine captains and coastguard officers, hundreds of miners, &c.—a general recognition and shaking of hands takes place, invitations and compliments are heard on every hand; arrangements for parties are made by young people during the ensuing week, so as not to entrench on each others engagements, or monopolise all the fiddlers. The older portion crowd the bar room of the Star, and other hotels, where the hosts and their wives are as busy as they can desire. The houses are full, notwithstanding the bed rooms, brew houses, and every available spot have been emptied to accommodate the swarming guests. "First come, first served, and money down," is the order of the evening, as the uncouth, red-faced, heated "tenders" (waiters) hurry to and fro with the foaming tankards of ale and glasses of toddy. The bar rooms are squeezed with receptacles, and tobacco is smoked until it is a matter of impossibility to see from one end of the table to the other: this, too, is termed enjoying themselves. This scene continues until ten or eleven o'clock, when every road, lane, highway, and field-crossing is thronged by parties returning, laden with as much provision of some sort as they can carry. The quantity of good things of this world conveyed from St. Just and Penzance for this occasion is really astonishing. Few people are seen intoxicated on the Saturday, as that would spoil the game; they, therefore, retire rather earlier than ordinarily.

On the Sunday morning, everybody that can be spared from cooking is at the parish church or chapel, in their bit of best. Many a new garment is purchased for this occasion that would not else have been bought, and many a child is well clad that would other-wise probably have been in tatters. A joyful sight, indeed, it is to witness so many happy, contented, neat, and comfortably clad labouring poor. Such a sight no country else in the world, with their sunny skies or golden sands, can display; not one of them are equal to an English village Sunday morning on such occasions. After service (the sermons from the clergymen being generally suitable admonitions for the occasion to the congregation), they retire to their homes, where tables literally groaning beneath the weight of

eatables and drinkables are provided in profusion. St. Just hospitality
seems never to be satisfied unless you cram yourselves to repletion.
They endeavour to show the warmth of their invitation, and hearti-
ness of welcome, by their importunities and anxieties for your
stuffing yourselves—a rustic but hearty way of demonstration.
After duly discussing the good things, "custom" (that is, "moonshine") is produced and imbibed *ad libitum—quantum sufficit* by all,
and a little extra by some, together with tobacco, as a matter of
course. The old people talk of bye-gone experience; what Captain
Boyns said and did at St. Just Feast some score of years before; what
Capt. Grenfell ought to have done at Botallack, when uncle
Nicholas Tremewen died, and what riches he left, how many ankers
of brandy he smuggled and drunk in his time, how badly Capt. ——
behaved to his wife, and what a wretch she was; the probability of
Wm. Chonnal marrying Susan Clark—in short, all the village
gossip; they tell their funniest tales, and drink each others good
health half-a-dozen times over.

After tea (about which the old men don't care), the young people
as a matter of course, go to chapel. There could be no feast at
church town without going to chapel—at least, it would be so
declared by the female portion of the community. On entering this
spacious building, the mass of not only well but absolutely smartly-
dressed people, who rise to sing "Praise God from whom all
blessings flow," is astonishing; the mind is puzzled to guess where
all the people come from. The first idea is that underground must
be as much populated as surface, for it seems impossible that the
apparently few scattered cottages can contain so many gay folks—
forgetting the great influx of strangers. Then peals forth a melody
to the magnificence of which we have before alluded; but, on this
occasion, when all the singers are present, a fine performance
ensues. A devout attention marks the service, particularly the
sermon. The dismissal takes place amid the utmost decorum, and in
ten minutes half the congregation are in the public houses, singing
the hymns over again. After a short stay they wend their way home
to supper.

The "Feasten Monday" is verily a holiday—save for the engine-
men on the mines, who must attend to their duties. The young men,
in the morning, amuse themselves with athletic sports, such as
wrestling, racing, and pitch and toss (cockfighting, badger baiting,
and such sports, to their honour, now are unknown). After dinner,
all hands—young, old, halt, lame, and blind—repair to church

town, where every kind of vehicle, from the gay Penzance post chaise to the humble Sennen sand cart, may be found having brought its crowded quota of visitors. There, too, may be found Mr. Punch in full play, amusing the crowd of gaping children and rusties with his tragio-comic, but noisy, performances. In juxta position, is the fanatical, enthusiastic, temperance advocate, hurling his anathemas, both here and hereafter, on the devoted heads of all who differ from his doctrine, consigning, with the most profound self-complacency, the whole host of brewers, distillers, landlords, and tipplers to a dreadful fate, at which pastime he labours in vain, on this day at least. There, too, are the players and minor exhibitions of "Truro Fair," with many of their camp followers, who, in St. Just, are obliged to keep civil tongues in their heads, and restrain their priggish tendencies within bounds; the "chickens" acting like one man, any robbing or swindling of the humblest mining lad would be the signal for the annihilation of the offenders. These have a wholesome dread of shafts and pick-hilts, remembering the circumstances of poor Laurence's Pavilion, the performers in which wanting a skull for Hamlet procured one from the churchyard. Unfortunately for them, it was rumoured about as belonging to some parishioner; the poor strollers had to precipitately retreat, or they would have been roughly handled by the excited miners, who hold the memory of their deceased in the greatest sanctity.

The public houses swarm with visitors: fiddling, dancing, and revelry may be heard on every side—enjoyed by the miners, their wives, sweethearts, brothers, and sisters—all is gaiety and boisterous good humour, fights seldom resulting; or if any, at an advanced hour. These scenes form the amusement of the lower orders: cribbage, whist, the pipe, and "moonshine", at their own houses, form that of the elder and substantial portion. "Pope Joan," or some other round game at cards, succeeded by country dancing, "blind man's buff," or "forfeits," that of the younger people, who all keep it up until the small hours chime.

At church town (next morning), the numerous sodden, yawning countenances attest overnight scenes we will not display or criticise; whilst demands for soda water and brandy, or gin and bitters, evidence what their betters had been doing. Perhaps a few glasses extra were enjoyed. The scenes of the past evening are enacted over again by all hands on this; on the next, the savings of the six months are reduced to mere dribblets, the circles of topers to a few well-seasoned, hard-headed, well-to-do personages, who recount the pleasures of the future time, and, at parting, congratulate each other on being still able to see the young ones out. The week over, little more is thought of the festival until six months have rolled away, when the same provision takes place. "Moonshine" must be, and is, provided. The same joy and festivity takes place under the same circumstances. Long may such scenes and bonds of harmony and good fellowship be continued, for, indeed, "It is a faint heart that never rejoices."


## THE MINER'S COURTSHIP

LOVE, THE GRAND LINK TO, AND LIKENESS OF, THE Godhead implanted man—the wonderful principle of our nature, so much canvassed, so little understood, so applauded, so prostituted —so wonderfully seated in the human breast, develops itself in humble society as forcibly and truly as in the more refined and artificial: we hope to portray one example in proof.

Poets and painters have long delighted to embody their ideas of pure, unsophisticated love, as Colin and Lavinia, in Arcadian bowers, or as gentle shepherds and innocent swains, with blushing nymphs and rosy milkmaids; and the stage, to rudely caricature the finest principle of human nature, as a red headed, clownish booby and smart winning chambermaid or cosey barmaid; both pictures, we think, are equally unnatural and overdrawn. Men are but men in all spheres of life, and when more is claimed for them, or when more is assumed than nature dictates, hypocrisy is practiced. The passion which exerts such powerful influences and promptings in our nature is as erratic and singular in the selection of its objects as its sources are mysterious; and, when once truly formed, it becomes absolutely uncontrollable, either resulting in ecstatic, lasting enjoyment of its purest principles, or in the destruction of that Reason which probably gave it birth. If we derive our picture from nature, we surely cannot err; we will, therefore, try the effect of our photographic lens on the subject. The names of the *dramatis personæ* in our illustrations we, for obvious reasons, conceal. To many persons the circumstances are partially known, and will be recognised; to them we refer for the faithfulness of our portraits. A youth, son of a well-known Cornish mine captain, who was learning his profession as his father did, and as all should acquire it—*i.e.*, by practical experience and hard work—in the course of his peregrinations, saw a young lady for whom, though in a somewhat superior grade of society to himself, he felt deeply interested at first sight: he knew not why; in vain did he try to suppress the impulse her person had caused—her form being continually present to his imagination; in vain did he reason to himself on the utter hopelessness of his heart's aspirations; in vain did he seek to hide from himself the deep rooted affection she had already implanted; hopeless appeared the chance of his ever obtaining an introduction to the object of his devoted admiration,—she was the village belle. The circumstances of their parents were not so dissimilar as to make a wide distinction of their families' future prospects, still they were not on terms of intimacy or acquaintance. The modest youth, however, could never persuade himself so sweet, so fair, so gentle a creature would condescend to give her society, her hand, or her heart to one in apparently so humble a condition. We are not so poetically inclined as to describe our hero and heroine as Venus and Adonis, but we must do them the justice to say that they were each considered models of their sex. He, from his athletic daily exercise, possessed a manly form and gait; she, from her youth, feminine pursuits, and exquisite beauty, became, as we have before said, the belle of the village.

The secret passion had preyed on William's heart for upwards of one year, when chance gave him the opportunity he had so long and anxiously desired. A local institution, for encouraging the arts and sciences, had held a public exhibition of the inventions and works of art sent in by competitors, at which William, for an important improvement in mining machinery, had obtained considerable distinction, his invention displaying meritorious ability; he was invited to spend a short time at a gentleman's house in the neighbourhood, at which Elizabeth was also a guest. How his heart throbbed when, for the first time, he found himself under the same roof, and on an equality with her! Aspirations before unfelt and unknown now took entire possession of his soul. Hope, that heavenly messenger to man, whispered to him, and he listened to her suggestions with ecstacy; that he might by endeavour yet raise himself to a proper grade became his conviction and determination. What were his emotions when she congratulated him on his eminent success, and expressed her approbation of his ability? The poor youth had almost forgotten his natural serenity of mind and habitual reserve; he was too much rejoiced to enjoy his happiness; his cup was filled to overflowing. It is not surprising that a youth who had

so prominently distinguished himself as to draw down high en-
comiums from the President in his speech should obtain a considerable degree of the ladies' society and conversation. Whilst
discoursing with her, his beloved (for now he felt and acknowledged to himself she was such), his brain burned, and he almost
longed to be delivered from such a vortex of delirium, and to have
time in secret to ponder over his situation. To him she was all
kindness—the very embodiment of angelic forms and sympathies;
every word went as an arrow to his soul; where they became fixed
for ever in the wound they made. He retired; and in a degree of
frenzy sought his pillow, but no rest came to him; sleep had refused
its duty to close his eyelids, and bring that rest so necessary to quiet
his disturbed mind, that day's accumulated success, joy, and happiness being nearly too much for him. Such is the mind of man—so
nearly is the balance of our reason adjusted, that a small excess
becomes as fatal to its action as overwork to any of man's ingenious
contrivances. At length Hope again came to his assistance. She was
the only one of the Triad spirits who had charms for him: she
calmed his troubled soul and brought her sister Faith to his assistance, inspiring that self confidence so admirable in man if it be not
allowed to dwindle into egotism.

Soon afterwards, returning from his employment in his working
costume, he saw Elizabeth and her father coming towards him on
the same path; on their meeting, the former addressed him familiarly, and the latter invited him to join them in a walk to their
house, at the same time questioning him on some particulars
relative to his invention, which had, by this time, obtained some
celebrity. After a few apologetic remarks he yielded a willing
consent. William felt considerably less embarrassed than he had
expected—he felt the influence of Hope and Faith on his soul, and
his confidence never forsook him. He now looked on her as his, and
from that moment a fixed determination to deserve and to win her
affections became the sole idea and end of his existence. At church,
on the following Sunday, opportunity of recognition again offered,
and then lovers only know how,—whether there be the subtle
influence among mankind called by the German philosophers
"odin" we know not, but we do know lovers meet accidentally; on
which occasion William perceived, by the conscious blush, the
half-suppressed smile, and the tell-tale index of the soul—the eye,
that his presence was anything but disagreeable to Elizabeth.

At length, with a manlike courage, that did him honour—

without breathing a syllable of his smouldering passion to her, and whilst she was on a visit—he waited on her parents, and frankly declared his now settled passion, and determination to render himself worthy their esteem and daughter's love, if he dared hope for a return—stated his consciousness of their different positions in life, which difficulty and obstacle he had resolved to overcome. He pleaded with an earnestness and eloquence the most fervid love could only have prompted, and waited with nervous anxiety the result, asking permission to be allowed to call again in a few days.

The parents of Elizabeth, although taken by surprise at the presumption of the young miner, could not but admire the disingenuousness of his conduct and openness of manner—a proceeding which the father highly approved of, as he said it could only have been dictated and prompted by the purest motives. On Elizabeth's return, her father sent for the youth and in his presence told his daughter of the interview he had given him; and now a scene occurred such as the pencils of Bulwer or D'Israeli could probably do justice to; but, as ours is only accustomed to ruder subjects, we prefer leaving it to the reader's imagination. Suffice it to say, William left the house encouraged, and resolved at once to effect the object of his ambition or perish. He felt the flame was exhausting him, and hastened to raise himself to opulence, ere he obtained her, preferring to be the victim of his gnawing passion rather than expose her to ridicule or discomfort by her connection with a common miner, as he appeared to be.

After a few months, he settled to go to Australia for an adventure at the then newly-discovered gold regions, where he flattered himself his mining knowledge would be of good advantage. Neither the wishes of his own and her parents, or her tears, could divert him from his purpose; after a most affectionate farewell, the meritorious youth sailed from Liverpool for the scene of his golden dreams, where he arrived in safety; but the seed of his fate had been sown, though he knew it not. He repaired at once to the diggings, where, as he had expected, his profession stood him in good stead. Strange to say he was singularly fortunate, but so much the worse for him. How often is apparent good but a deceptive reality! how frequently too, we mistake the shadow for the substance! His success excited him to over-exertion and anxiety to secure his glittering prize. These fostered the worm already praying on his vitals, but he heeded it not—gold, gold, gold was, save Elizabeth, his only idea—the only things in the world that had charms for him, and both apparently his. Comforts he cared not for, sympathies he sought not, health he disregarded; all these would be his on his return. Vain hope; deceptive vision! it was too late.

As in all cases where gold becomes the besetting god of existence, gold came; he was successful beyond precedent. In a brief period he had realised a sum, half the amount of which would have originally satisfied him, but now as long as gold was obtainable there was he; regardless of sultry sun or midnight dews, he heedlessly wrought, until the place become less productive. Fatal delay! He then wrote home to say he had been abundantly blessed by Providence, and that all he desired would now soon be within his reach. Poor fellow, little was he aware of the fatal drop in his cup of joy, little did he suspect the asp beneath the leaf of the beauteous flower! During the continual excitement and exortion consequent on his gold washing employment, he did not perceive he was working nature up to a pitch that must necessarily have a powerful reaction; he was too much engaged to feel it. When on board the ——, on his homeward voyage, the usual sickness and monotony of a shipboard life, to-gether with his thoughts continually racking on his reception by Elizabeth and her parents on his arrival, nurtured that reaction he ought carefully to have guarded against. At length it burst forth in the most aggravated form of wild delirium and prostration of strength, both bodily and mental. The medical attendant at once saw the case was past human skill, and utterly hopeless; a few days illness released the poor sufferer alike from his hopes, his joys, and his pains. It is to be hoped at his death the Triad Angel was in attendance, to waft his innocent soul to its haven; at all events, let its influence rest with us to let it be so. The body of that exemplary youth, then in its twenty-fourth year, was duly consigned to the deep; the chests and chattels of the deceased were sealed by the captain, in the presence of the chief officers, and removed to a place of security. The splendid steamer careered as swiftly and gaily as ever over the bosom of the deep, and on her arrival in port the death of the miner had been nearly forgotten.

Meantime, William had sent advices of his splendid success, of his ardent hopes, of his little mementoes in the diggings. Carefully arranged in his private desk, evidently to decorate his bride, were some most exquisite specimens of gold and quartz (the author has seen), set in the country, as rings, brooches, &c. He also enclosed remittances to purchase a neat little cottage they in their rambles had often admired, but of which his fondest dreams had never led him to hope to become the proprietor. He had also in his letters to her expressed a wish she should be at Liverpool, to receive him on landing. This, as may be supposed, was cheerfully complied with; having a sister married and resident in that town, the maiden, now full of joy, and with the entire approval of her parents, hurried off to that town, in readiness to welcome her now proved to be devoted lover. Every possible and suitable preparation had been made for the celebration of the nuptials, and on the very day the —— steamer was due the telegraph gave notice of her arrival off Holyhead. A river steamer was hired for the occasion, and the joyous, gay, expectant, beauteous bride, in all the vigour of hope, health, ecstatic happiness, embarked with her friends, and were soon on the deck of the magnificent ship. The steward was immediately sent for, and eagerly asked where William H—— was. For humanity's sake, it is to be hoped, the blockhead, replied, "We had one on board, but we threw him overboard at the Cape, and the skipper has his boxes aft!"

Our pen (as Sterne says) we govern not it, in describing the scene on that deck; it wavers, hesitates, and falls, as did she fall; indeed, we know of no author we can quote to do justice to it—we feel our utter incapacity. Thompson, in his beautiful example of Celadon and Amelia, where the latter was struck by a thunderbolt, says, "But who shall paint the lover as he stood?" was but a faint apostrophe. One of our poets speaks of the "ecstacy of woe", if there be such a feeling (God grant there may), that was the culmination which hurled Reason from her seat, and in one moment transformed that God's last, best work from all its beatitude and magnificence, as his image, into a living corpse—a drivelling idiot, which to this day she remains—a monument of blighted affection, of the wonderful principle called Love, and a solemn and awful warning that it, the last greatest gift, the reflection and part of His own essence granted to man, should not be tampered or trifled with. Here will we hold—the good will pause with us, and shed a tear; if soft humanity e'er touched your breast it will be no disgrace. Sterne's angel still keeps the record; then, if you identify and feel the lover's sorrow, oh! reader, bedew thy cheek; it is a holy tribute, and will do thy soul good—try!

In His inscrutable wisdom the Almighty, for some wise end, no doubt, suffered this awful visitation! Great God! Omnipotent! Jehovah! if ever it be thy command we are silent; thy high behest what finite dare anticipate? Not a sigh or groan was uttered; a blank was the mind that a moment before had been so vigorous—a wretched living corpse that fair form! Reflect, oh! man; pause and consider when that mysterious golden link is broken that thou art not "a *little* lower than the angels;" therefore, trifle not with that which is above all price, to which rubies are but glittering baubles! Our picture is done; the light has been long enough to transfer it. If it does not please we cannot help it; we know it is true—too true; we sympathise. Adieu! adieu!

We should append that the captain, his father, refused to touch the sum forwarded to her, to claim the cottage, or to have the little mementos disturbed from the scrupulously careful order in which they had been deposited for her.


## THE MINE

TO SEE A CORNISH MINE CONDUCTED AS IT SHOULD BE IS not an every-day occurrence, though nearly all in this country, as elsewhere, profess to be the very *beaux ideal* of that "consummation devoutly to be wished." Undoubtedly, there are many such —it would be invidious to particularise; in doing so, we feel that we should be committing an injustice to some; we shall, therefore, speak of one *incog*. As it embodies all the main peculiarities, we adopt it as the type of what a Cornish mine should be and is: many will know the portrait.

We hold it to be an essential that the employed be cared for as well as the proprietary; that the directors, or committee and agents, are the mediators between these parties; that whilst doing their duty they should remember justice is, or should be, even-handed—the success of the undertaking depending much on the due execution of these premises.

Accompanied by our tourist (from whom we had for some time parted company), we arrive on the mine, and at once go to the counting-house and enquire for the captain, to whom we introduce our friends; and, readers, it will be unnecessary to describe him further than to refer to our Photograph of "The Captain" for his portrait. He familiarly offers every facility, and, as none of his duties prevent him, he kindly volunteers to accompany us over the mine. "Do you intend to go underground, gentlemen?" "If you please, Captain." "Well then; here, Mary," addressing the counthouse woman, "get out and air three suits in three minutes. Let me see—we shall be worst off for shoes; let me see, have you large feet. Dear me! dear me! what shall we do for small underground shoes. You can wear mine—I can get a pair of the men's; mine have got good sound soles, and will be safer for you to travel with. Were you, gentlemen, ever underground before? Mr. H——, there, is an old stager, and can manage well enough." On being answered they never had had that pleasure, the Captain replied, "Well, Mr. H—— has brought you to a right mine; you have something to see here. Now for hat-caps; there's mine will fit you, Sir, and our second captain's will fit you, Mr. H——, and I can shifty. Dear me, Mr. H——, how didn't you send word you were coming! Never mind, I suppose now we've got clothes we shall do. Sit down a few minutes." Through the wooden partition we heard the following colloquy, in *sotto voce:*—"Mary, what hav'aye got in the count-house?" "Nothing fit for gentry, Kappen." "Oh, never mind, send off the maid at once for a few pounds of beef steak, and a bit of fish, if she can get it; and some new bread, and tell her to go to my house for two bottles of wine and a bottle of brandy. There is some gin in the cupboard, isn't there, and a bottle or two of porter, left from last account?" "Yes, Sir." "These strangers ought to have something when they come up; and be sure and have plenty of hot water to wash. Let me see, we shan't go down till twelve, and we come up at three; have'n all ready by four, that'al do; have't all nice, now." Re-entering the room—"Now, gentlemen, please to walk this way; we had better see the surface first."

We go into the great engine-house, in which is a splendid 80-in. engine, where all the mysteries of arbors, parallel motions, packing segments, throttle and other valves, condensing and other apparatus, are explained, and we ascend to the bob-plat, where the mighty beam, 26 tons, moves "like a thing of life," lifting its enormous load with apparent ease and perfect silence. From this elevated situation we take a bird's-eye view of the whole mine. The better to comprehend our future underground peregrinations, the Captain points out the direction of the lodes and shafts. That is Broad's shaft; that's such-and-such and so-and-so's shafts. We next visit the capital steam-whim engine, and have its peculiarities and different construction pointed out. Then attention is called to its working, and tell-tale so ingeniously contrived. Then comes a disquisition on the merits and superiority of wire ropes—an improvement lately made in the application of skips and guides in shafts—over kibbles and chains, a subject so evidently exemplified that we could not

withhold our extreme approbation. Then the beautifully laid out
floors, with all its labyrinth of steam crushing-machines, buddles,
jigging machines, and improved slime-pits, where all the process,
from tin stone to the mysterious little brown sand called "black
tin," and from rocks of copper ore to the "dole" ready for sale,
were lucidly explained and illustrated. The party were now shown
the houses prepared for the workmen to get their meals in comfort:
a large room, with tables and cupboards for their stores; the room
with hot air to dry their wet underground clothes gradually instead
of roasting them before a fire; the comfortable and extensive hot
bath in which to wash themselves when coming up after their
labour; the women's room, where a cooking apparatus is provided,
in which they can warm their pasties or other victuals, heat or
prepare tea or coffee; and by a large pipe, filled with hot water
running through the apartment, they can warm their feet whilst
enjoying their meals, with every comfort they can desire—an
admirable arrangement; it prevents that wandering about and idle
conversation, that mixing of boys and girls in play-time, so much to
be desired; and when the bell rings they are on the spot. No wonder
the best miners and work-people flock to such mines as these! But
the Captain says we must make haste, and we must obey him, and
go to the counting-house, where the underground clothes are now
waiting, when he opens a drawer and says, "Whatever valuables
you have, gentlemen, please to put in here, and they'll be safe." He
then hands the "suits" to the strangers and retires. In a few minutes,
after a little trouble about waist straps and braces, the "strangers"
stare at each other in wild amaze to see what figures they present in
the strange costume of flannels and duck. The Captain now re-
appears: but, oh! how "changed" is he by his change of garb: a
dress of one of the miners—his being used by the visitors—red as
ruddle, ragged, in tatters, with about a pound of captain's candles
in his hand, and enquires, "Are you ready, gentlemen? How deep
do you mean to go, Mr. H——? We have a good ladder-road.
We've got a good course of ore in the back of the 60. We must not
go further than we can do good for strangers. Our adit is 33 fms.
from surface, and 60 more makes 93 fms.; but if you can go to the
80 or 90 there is something worth seeing, and I should like Mr. H——
to see our north lode. Well, we shall see how we get on; we'll go
down the footway shaft first. I'll go before with one gentleman,
and you, Mr. H——, go down before the other. Take care and hold
fast by your hands, and never mind your feet. Here's some clay; I'll

fit your candle, and Mr. H—— will fit the other gentleman's." The strangers now approach the shaft's mouth—not without a vague dread, and heart pit-a-pat, at the novelty and supposed danger of the feat, overpowering the captain with thanks and compliments for his politeness.

At length the descent is commenced, and the first ladder to the plat got over very well, when the candles are lighted, and the descent begins in good earnest. A little caution is used, the captain calling out, "Now, if I travel too fast for you, tell me, and I will go slower." The first "sollar" (resting place in the shaft) is now reached at about 10 fms. They then turn round on to another ladder, and this process is repeated until they arrive at a large excavation. "This, gentlemen, is the adit level, where we empty the water pumped from below by the great engine you saw; this tunnel extends more than $2\frac{1}{2}$ miles, and it is the lowest level we could procure above water. All our deeper levels are reckoned from this—we can see the back of the lode here. Walk this way; but first let us trim your candles."

By this time the strangers begin to perspire rapidly, the pearly drops standing on their brows. The Captain encourages them by saying, "We haven't got much more hard work now—the shaft is on the underlay" (in a sloping direction, the same as the vein). After walking 200 yards in the adit through water and mud, the "lode" is reached, and its walls, capels, gossans, &c., explained. No mineral of value is in it, but its peculiar indications are pointed out. They then go down ladder after ladder, as before, apparently without end, resting every 10 fms. to draw breath. At length the 80 fm. level is reached, the party perspiring at every pore, and gasping for breath. On examining the thermometer, it is found at 86°, and they clothed in flannel and duck from head to foot! They now walk about 40 fms., when suddenly "Bang, bang!" resounds through the cavern, making our friends jump and exhibit signs of terror. On being assured that there is no danger, the party proceed to where the men are at work. Poor fellows—three men, and one little boy not 10 years of age, working in this cavern and atmosphere for eight hours!

The Captain addresses them in their own patois, thus—"Well, my sons, how's she looking now?" "Don'naw, Kappen." "Who shud, then, if you don't?" "Don'naw, Kappen; torrable bluv. Come and see for yourself." Upon which the captain carefully examines the lodes for some minutes in silence, and then says, "Here's a branch gone off here to the left; take care what you are about—I'm afraid we've come to the 'nose of a horse'" (a peculiar mining term, meaning an isolated piece of rock in the lode). "I reckon we are, Kappen." "How didn't ye come up and report this to me before? Didn't Kappen John (the second captain) see un?" "No, or else you wud have nowed it, I suppose. I han't a seed un five minutes." "He's surely split: just examine this end, Mr. H——. What do you think of it?" We replied, it was evidently so. The Captain then admonished the miner—"Now, you take care what you are about. I must come down again to-morrow, and if you notice any change, send up for me, and I'll come down to once. Es there anybody up in Wylliams's stopes?" "Iss." "How is she looking" "Very well, I reckon, Kappen. I suppose they got a bra pile of work up there." "Do'e think these here gentlemen cud git up there to see un?" "Iss: there's a good road anuff."

The strangers were then shown a hole, very like going up a chimney, with pieces of wood placed here and there, to be enabled to climb to the "stopes", about 40 feet above, and perpendicular, the slightest slip ensuring a broken limb. At length the task is attempted, the Captain sending a miner before, to prevent any stones falling, when he himself leads the way. "Give me your hand, Sir: don't be afraid." A miner comes behind, laughing and enjoying the fun of the terror of the strangers, who, after incredible labour, reach the "stopes," and a place like a large oven is entered—here are some of the richest of the mine. "'Tes like a jeweller's shop, essen a', Sirs?" exclaims the miner: "I ben 22 year underground, and never seed nothing liken afore. Es a beauty, essen 'a, Kappen? 'Tes too good for to eat, seeming to me"—as it did to the strangers, it being red oxide and carbonate of copper (malachite). The Captain with pride takes up a pick, and says, "Now, gentlemen, 'tis not every mine in England can show an end and back like that, I assure you. Can they, Mr. H——? You have seen a great many, did you ever see a better?" On being answered in the negative, specimens were broken by the delighted though somewhat terrified strangers for their cabinets. The lode being "desued" (denuded of its capel), the Captain cried, "Stand back; give me a pick and gad, and I'll take down a bit or two," when he took out two pieces, each weighing about 1 cwt. "Here, gentlemen, that stuff will yield 60 per cent as it is broken: these two pieces are worth 5l. or 6l. as they lie. That's the way we make the mines pay when we have a good lode." Then addressing the men, "You'll make rare wages this month, my sons." "Well, Kappen, we only got 40s. a man, and nothing for the boys, last; and if it hadn't ben for we, you nor these gentlemen, nor the 'venturers nuthur, wud never have seen these backs. Besides, 'tes killing work, the air is so bad"—a fact displayed to demonstration by three of the candles going out.

All then descend in safety as quickly as possible, the only injury being grazed knuckles and shins. A walk back to the shaft succeeded, and the ladders again descended to the 90, where the lode was again found, and the same ceremony repeated.

"Now, gentlemen, I see you sweat rarely"—and on examining the thermometer it is found to be 108°, the strangers expressing their surprise at there being so little water, which was explained as being far below their feet.—"Now, gentlemen, I've got a doctor with me; try his medicine," producing a small flask of weak brandy and water—a welcome administration, indeed. "Now, this is deep enough for you, so sit down and rest yourselves. You are more than 100 fms. from the surface, which is something for you to say when you get to London. I want Mr. H—— to see something in the 114; but it is a bad, wet road, and will hardly do for you. We won't be long. I've got a cigar for you (eagerly accepted), and by the time you've smoked it and rested yourselves, we shall be up again." On our reappearance, we found our visitors now considerably re-freshed, and surrounded by a group of miners. They requested to be again introduced to the doctor, which having been accorded, the works in the 90 examined and explained, and a hole or two fired to show how it was done, the ascent was commenced in right earnest.

Arrived at the first 10 fms., or 80 fm. level, a walk of about a quarter of a mile along it brought them to the cross-cut. "You are now under the big shaft I pointed out at the surface." We leave this lode, turn to our right, and steer north to find the north lode—as inexplicable to the strangers as the parallax of the stars. After steering north nearly another quarter of a mile, the north lode was reached, and found very rich for copper. This was examined as before, and a retreat made. Another walk of a quarter of a mile brought them to the engine shaft. "You are now right under the small house at surface, but the engine is away many fathoms from us, by the underlay of the lode."

They now commence to ascend again, when, the ponderous balance-bob at the bottom, the wonders of hydraulic machinery, and mode of fixing, perhaps, having been explained and shown, and all the plunger poles, H-pieces, doors, clacks, and charging boxes (such incomprehensible and expensive items in cost-sheets), fully dilated on, and the whole system of casing and dividing shafts exemplified, the tedious process of ascending being carried on at intervals during these valuable instructions by the Captain to the now panting and sweating half-miners. At length a gleam of day-light appears—"There, gentlemen, there's daylight;" but no response comes from the now thoroughly exhausted party.

On reaching the surface, "Thank God that's over for once!" burst from one gentleman, and "If ever they catch me down that hole again, I'll be ——," from the other. The doctor was again applied to, and in five minutes all were right. Fun now became the order of the day:—"I would give a guinea if your wife could see you in your costume."—"And I would give a like sum if Miss —— could have your picture. Why, you look like a man now." All hands repair to the count-house, where orders had been obeyed—everything was ready. A comfortable wash and change of clothing was soon effected, and the party were as well as ever, save a little trembling of hands, caused by over-exertion and climbing too quickly.

The Captain now introduced the purser, the sampler, and a friend, who were assured by him that he had never seen two gentlemen behave better underground. The count-house woman now warns them that dinner is on table up-stairs, and the now large party, nothing loth, partake of the substantial beefsteak dinner, the captain occupying the chair. A glass of grog was afterwards en-joyed, the strangers thanking the Captain over and over again, assuring him of their entire satisfaction at everything they had seen, and that the day's experience had given them more information than they could have conceived; that they could now see where their money had been expended, and why mining was so costly; and felt convinced that every adventurer, before grumbling, should do as they had done—have a day on and under a mine, to be convinced of what mines and miners really are, when properly managed—an opinion which, we hope, will be responded to by our readers, who, if they will try the experiment, will fall into the same ideas as our friends, who, when at home some time after, said—"Go and see them: they are fine honest fellows. They and their mines deserve en-couragement, and have been grossly misrepresented and maligned."

## The Triumph

PRIDE IS A PRINCIPLE OF OUR NATURE, THAT EITHER becomes a vice or a virtue in its possessor, according to the culti-vation it receives. If it be allowed to grow unchecked, it bears the evil of vanity, but by proper restraint it blossoms with the cardinal virtue of self-confidence; the one prompts to heroic deeds and great accomplishments, the other to egotism and egregious follies. A remarkable instance of the just application of this principle lately fell under our notice, on a visit to the Copper Ticketing held at Pool, on Thursday, Oct. 1 inst., where a "triumph" was achieved, by the captain of West Seton Mine filling the chair on that occasion for the first time. True, he was a proud man, and had the manliness to acknowledge it; in doing so, he enunciated so many truths and most excellent mining maxims, that we think a "Photograph" of the entire meeting will be appropriate to our series.

It will be unnecessary to say more of the ticketing than that it is a meeting at which the parcels of ores from the mines are tendered for by ticket, the agents of the smelting and mining companies being present. A large amount of business is transacted, generally about 25,000*l*. After the business of the day, a splendid dinner is enjoyed by the parties in attendance, wholly composed of the agents of the companies, invited guests, and "strangers," by which term is understood gentlemen non-resident in the counties of Devon and Cornwall, who are always welcome. The representative of the mine selling the largest quantity of ore takes the chair: as such is a good reason for rejoicing at any time, the mine first doing so is generally hailed with acclamation, and "heading the list" is complimented by a present from the adventurers of a round or two of champagne, which on this occasion was liberally afforded by West Seton; well could they afford it, their parcel of ore being 502 tons, and the amount realised 3806*l*. 2s. 6d. Being in the neighbourhood on a tour of mine inspection, we received a polite invitation, which we, of course, accepted. After the conclusion of the business of the day, and the cloth being drawn, about 30 or 40 gentlemen being present, including two strangers, besides ourselves, the prospects of the various mining interests, the prices for copper and tin ores, the improvements in machinery, the various modes of dressing, the dues of landlords, &c., were freely discussed, as well as a handsome dessert and ample supply of wine, under the influences of which, and the presence of so much ability, a conversation mutually beneficial and instructive was enjoyed by all for two or three hours. These meetings taking place almost weekly by the agents of the different districts (for all mines are not able to sell ores regularly every month), creates an inter-change of good feeling and ideas, of the highest importance to the welfare of mining. We were really delighted to be recognised, even at this western ticketing, by "east country captains." This fusion of districts must and does tend to good.

After the loyal toasts, "Lords of mines," "Fish, tin, and copper," came "Success to West Seton." The Chairman then rose and recorded a series of facts it is the object of our paper to record, for the benefit of wavering, timid shareholders, therefore for mining generally. He commenced by saying he felt proud that day, for he had achieved a "triumph," the object of his hope's fondest aspirations; that was, not for himself, but for his mine, on the adventurers' account, to be placed in so honourable a position as to occupy that chair, and be at the "head of the list." He could assure them it had not been achieved but by long and anxious perseverance and a determination to overcome all difficulties, by which they had triumphed. They had been upwards of 14 years in accomplishing it, but having at last reached the goal of their desires, he must acknowledge, even to himself, they and he had reasons to be proud. He said they individually afforded instruction to adventurers who were not at once prosperous not to despair; had they done so in their earlier day, during their period of trial (for such they had), their splendid mine, now at the head of the list, would have been lost to them; that they by perseverance were now in a position to repay all anxieties, and to fulfil their most sanguine expectations. Mining, he thought, had been improperly represented, by parties holding out hopes of immediate returns which were seldom if ever verified; but if perseverance such as they had in them were carried out, disappointment seldom occurred. He quoted a few statistics of that day's proceedings, which he thought would be of some value, if generally known; they had that day sold 3935 tons of copper ore, the produce of 23 mines, for the sum of 25,033l. 14s., out of which no less than 12,000l. to 14,000l. would go as profit into the hands of the adventurers, a result which he thought was a convincing proof that mining was undoubtedly of vast importance to the community, particularly to the districts in which the metals are found, which are usually bleak and otherwise valueless. He felt perfectly convinced in his own mind that mining, if properly and honestly conducted, was and is one of the most lucrative and attractive pursuits of this great country; but it could only be by perseverance and the outlay of capital they could ever hope to fill that chair, and achieve a "triumph," which he hoped they would all by-and-bye enjoy, as they had.

"Better prices, and a better standard," and numerous other toasts, having been honoured and replied to by gentlemen present, that of "Strangers" was cordially received; on which they stated their thorough appreciation of the compliment. Having been over some of the mines during the last few days, they could not resist the temptation of being present to witness the most pleasing termination of the routine of mining procedure—the sale of the ores, and the meeting of so many gentlemen of such sterling worth and ability. They could assure them they should leave Cornwall with very different impressions of its mines and miners than they had previously entertained; everywhere had they witnessed the utmost attention from the lowest individual employed, and had admired some most astonishing efforts of human ingenuity and industry,—they had seen everything to admire and nothing to condemn. That the observation of the Chairman was just—without perseverance nothing could be successful. They thanked all present, but more particularly the Chairman, for his valuable and excellent speech, and to ensure them, though they had travelled thousands of miles, and only just returned from Australia, they had seen nothing to give them more sincere pleasure than their visit to Cornwall and its mines. We should not omit to state, that we felt pride on being recognised as the author of these papers, which were stated to have afforded much gratification and amusement to the readers of the *Mining Journal*,—that they were calculated to do, and had done, much good, by representing Cornish character and ability as they existed; and was requested to continue the series, which would undoubtedly still further assist in the great work. We expressed our thanks, on behalf of the proprietors of the *Mining Journal*, assuring the company it was their anxious wish to do all in their power to forward mining interests, by disseminating truths and facts as they could obtain them. That as these papers appeared so welcome, and were so much approved, they should be continued, and that that day's proceedings, we thought, would make a capital subject for a Photograph, to be called the "Triumph," which had that day been so gracefully acknowledged and displayed by the Chairman.

Now, oh, reader! this matter was not done in the dark; open day was the means by which the picture was transferred and secured. Such is a faithful representation of a Ticketing, and the characters attending it. There is, indeed, much that, as the "strangers" said, was "just" in the Chairman's remarks, and so is there much to be gleaned from theirs, when they say, on witnessing Cornishmen and

mines as they really exist, they arrive at very different to precon-
ceived opinions. As it is impossible all can visit Cornwall to see for
themselves, the object of these papers is to give faithful transcripts to
obtain the same desirable end, which, if we accomplish, we shall feel
a just pride, and confess to ourselves that we, too, have achieved a
"triumph."

## THE BAL BOY

NO APOTHEGM IN THE ENGLISH OR ANY OTHER LANGUAGE
is more trite or true than that "One half of the world knows not
how the other half lives." Little does the child "born with a silver
spoon in its mouth" heed or care for him that is born with a wooden
ladle. There always have been hewers of wood and drawers of
water, and will be to the end of time: it is one of the conditions of
our existence, but that does not alter the relative duties of the one
towards the other. If the Almighty in his providence has seen fit to
raise one star above another in glory, they and we have each our
relative duties to fulfil: it will surely be required of us how we have
managed the talents, and taken care of our trusts.

We are by no means maudlin philanthropists, or Exeter Hall
declaimers on the horrid enormities perpetrated on black slaves at
home and abroad; nor are we sentimental juveniles or piously inclined young ladies, so graphically depicted by our facetious Fleet-
street contemporary. No! we are made of sterner stuff, or we should
not have witnessed the scenes we are about to describe: we are of the
masculine tribe; still we deem sympathy with the weak and innocent no disgrace, but, on the contrary, a cardinal duty, the neglect
of which is a venality. Our tale is brief; we claim but a few minutes,
and the poor little objects shall speak for themselves, and declare
their own tale of woe.

The "bal boy" is the term usually given boys when they first go to
work on a mine: they are usually placed on the dressing-floors,
where they are taught to sort the various ores, so as to know them at
a moment's glance. On some of the larger mines, scores of boys and
girls may be seen at this employment, selecting the different metals
with surprising rapidity. Boys are then taught to buddle, jig, and
prepare the ores for market. In tin dressing, more particularly, they
are exposed to almost constant damp. In buddling, their bare feet,

and legs as far as their knee, are continually wet. Until lately but few mines had even sheds to cover them from the weather. They were then almost continually drenched to the skin during the winter and spring months. After being taught the dressing process they are usually sent underground—at the age of 14. To this we see no objection, at this age their persons being sufficiently matured; but we do and will protest against children being allowed to be sent to great depths at an earlier age.

Being ungound lately in many of the deep and extensive mines of Illogan, Gwennap, and Redruth, our heart ached to find so many poor children, of much earlier age, being deliberately murdered (we can use no milder term) to save or earn their parents a few shillings a week. At a depth of 145 fms., 870 ft., from surface, and up in a rise, a place very like an oven, and to gain which the miners and ourselves experienced considerable difficulty, we found a poor little pale, wretched child at work, when the following Cornish colloquy took place:—"How old is ta me son?" "How old am I, Sir? [a miner invariably echoes your first question] why I shall be ten year old come next birthday." "How long have ye worked underground?" "How long have I worked underground? lemme see, better than two year, and worked here all the time." "Who's thy father, then: what do they call'en?" "That's may fayther," pointing to an old man in the place. We then questioned the father as to his own age, which he said was 36, he believed, but he could not tell. He could not read or write; he went to chapel every Sunday, when it wasn't his turn to watch the mine. He appeared to be 60 at least, and we found he had been sent underground at ten years of age! On remonstrating with him for bringing a child of such tender years to such a tremendous depth, and to such an atmosphere, 98 degrees, and withal highly sulphurous and arsenical, from the peculiar minerals being raised, the father acknowledged he knew it was bad enough— "For," said he, "when I was a boy myself I have many times had a mind to lay myself down on a pile of ore and die, after a hard day's work, and then to climb 200 fms. to grass, loaded with borers; and I did once fall asleep in the levels, and when I waked my candle was out, and there I was, forced to wait till somebody came to see for me. But what can a body do now, Sir? Them buddling and trunking machines have near done away with boys at surface: besides, he saves me a pound a month." So, there was this poor child, created in God's image, sacrificed to the idol Gain, and for such a paltry sum!

We found this to be by no means an isolated case. In one place, at the bottom of a very wet shaft, was a poor little fellow, holding the borer whilst a youth about 18 or 20 was hammering away at it with the strength of a Cyclops. An error or false blow on his part would probably break the child's arm: the slightest neglect of duty of the poor little fellow would be sure to entail a box on the ear from his companion. We soon found from experience that this was a dreadfully cold and wet mine, ten minutes sufficing to saturate our clothing, penetrating to the very skin, though of far superior quality; whilst the rags in which the miners were arrayed afforded but a mockery of protection. On enquiring of the child how long he worked, he replied—"Six hours, Sir; and long enough, too, Sir." To which we yielded a candid and ready assent. Poor little fellow! six hours in such a situation, and then to have 110 fms., or 660 ft. (the depth he was working in) perpendicular to climb on ladders, the steps of which were 10 and 12 in. apart, to overcome which an amount of physical labour must be expended truly frightful to contemplate—known only to practical experience or surgical professors. Yet this is the fate of hundreds of children not more than eight years of age! No wonder the gravestones bear the early dates they do—no wonder miners' consumption is prevalent—no wonder ignorance prevails! After such a journey, and such work, can children be expected to give much attention to learning to read and write—can they be supposed to take that recreation so necessary to childhood and boyhood? Is it any wonder that they associate, as they do continually, with men, become premature in their tastes and habits, contracting many they can never again abandon—can it be a matter of surprise that we see boys idling away their time listlessly, with the horrid pipe stuck in their mouths?—No; certainly not. As long as childhood be permitted to be employed in such situations, so long will these evils prevail. It calls aloud for remedy, but there has been no one to advocate their cause. Mammon stands in the way, and he is an all-powerful potentate. Why should not the protection afforded the cotton-spinner be extended to the poor little "bal boy?"

The Pantheistic theology of the Greeks and Romans, if carefully studied, seems evidently to have been derived from the circumstances of mankind, and not designed to suit or improve them, as is Christianity, if its principles be carried out, and hence the evidence of its divine origin.

Had the ancients sought out gods to personify the principles involved in the practice just detailed, could they have selected more suitable types than their Plutus and Pluto? The former is the god which rules supreme in the minds of those who permit such atrocities to exist, and the latter the demon that prompts, and to whom these poor wretches are heedlessly consigned, so that they but help to support the golden throne of the former. We see that the allegorical representation is perfectly true to nature, still we should have thought the more sublime system would have removed the adoration still paid to such idols. It would have been supposed that in this land of Christianity, where charitable institutions (to its honour) exist almost without a number—where countless temples of the various sects arise—where towers and steeples bristle, the service of these gods would have been so far expunged as to have prevented the immolation of little innocents on their sordid altars. Shame on mankind—shame on their institutions, that such things are permitted, even by parents, whose necessities should not even be allowed to plead as an excuse. Our blood boils with indignation, and our heart bleeds with sympathy! They are as surely murdered, body and soul, by a quiet, subtle, unseen process, as the poor innocents who have been bayoneted by the cruel Hindoo. But because it is at home, and is gradual, it is not noticed: still it is not the less true. We do not mean to condemn or infer that child labour should not be employed—far from it; but let child labour be suited to childhood, and not exercised in a profession dangerous to the skilful and experienced; but cruelty itself supervened when children are compelled to go to such depths and such horrible holes.

Considered in a social point of view, can we be said to be doing our duty to ourselves and to society in allowing such things—can the rising generation be expected to be what they ought and should be—can they ever fulfil the high calling whereunto they are called? Can we, when called upon to give our great account, render a just one of having done our duty, neglecting this?

We are no orator, but had we the philanthrophy of a Wilberforce, poetry of a Cowper, persuasive eloquence of a Peel, or the power of a Palmerston, our charity would be excited, our page would be graced with harmony—our voice would be raised, and our power exerted—aye! all combined should never cease their action, until this foul spot should be wiped out for ever, and these poor children enjoy that blessing which is their natural birthright—free air and proper exercise during their tender years. Why should not these privileges be extended to mines as well as to factories? On enquiring of the village schoolmaster, I found him a good man, bitterly lamenting that he could not get the children to school; and even if they were sent their attendance was so irregular that he could not do any good. "It will never be any better, Sir, until it is stopped by law; for if parents can get a penny out of their children they will send them anywhere." We appeal to the humanity that has so successfully exerted itself in the cause of the sweep, which of the two is the greater evil,—the ascent of a few yards in a chimney, or the descent of hundreds of yards, and exposure to wet, with a foul atmosphere and hard work for six hours, then the frightful job of ascent, ere the poor child obtains his scanty meal and wretched bed? We think we hear it said—Can such things be? We answer, yes! we have painted but two as samples, thinking they are sufficient; they are but types, but they are facts patent to thousands. But use makes all things familiar, and use has rendered people as callous as eel-sellers. But eels have feelings, though they be skinned daily, and these poor bairns have bodies and souls!

And now, O ye fortunate scions of Nature! ye who were born with the "silver spoons," come forward—recollect your brethren of the "wooden ladles;" and thou, O fair goddess, Humanity (she is no idol)—we invoke ye all—come forward, assist in the good cause; your countenance alone will be of vast avail; come forward and show the world that, though "One half the world knows not how the other half lives," you have charity and Christian fellow-feeling, when you do know, to assist those who have no power to help themselves—to come forward and deliver them from their tribulation and oppression. Remember the words of your Divine Master, "Suffer little children to come unto me, and forbid them not, for of such is the kingdom of Heaven." Act on this precept!

## The Bal Seller

I HAVE LONG WAITED A FAVOURABLE OPPORTUNITY OF depicting an authentic specimen of this *genus homo*—a real, genuine "bal seller," such as exists more frequently in imagination than reality; one who offers a mine for the purposes of pure deception and robbery, and nothing else; whose whole argument is a "slocking stone," and whose sole ability consists of lying and unblushing impudence: such a man, in one instance alone, does infinitely more mischief than twenty able and good men can repair in twenty years. Fortunately, the facility with which localities can now be visited prevents or checks any such proceedings for any length of time, as proprietors and intending shareholders can easily ascertain, by themselves or their paid agents, whether the leading facts of the case are so or not. The *Mining Journal* now is the ægis of protection against such frauds; for were they now attempted to be foisted on the public, its columns, from some correspondent or other, would be sure to unmask the humbug, and expose the perpetrator. But to our subject—the "bal seller." As truth is sometimes libel, and as we have no wish to figure in a court of justice, like the subject of our paper, we shall use the author's privilege, and, whilst detailing facts, keep the initial incog.: to hundreds the subject is painfully patent, having been ruinous.

Well, then; on the estate of P—— is situated a mine, which was first wrought by intelligent captains, who executed extensive works, but having no less than five or six other mines in hand, at a time of great monetary derangement, found it necessary to discontinue operations; this was accordingly done, entailing a heavy loss. This was the very chance for the "bal seller," who proceeded as follows:—It should be mentioned he was the son of a Cornish mine captain of a very different turn of mind and disposition, but had settled in London for some time as a broker. Hearing of this mine, then called E—— S——, he called on a person at T——, whom he well knew had lost a considerable sum in the last adventure, and offered him a liberal share in a new company he undertook to form, provided he would give him an introduction to the proprietors. As a drowning man is willing to catch at a straw, so was this gentleman eager to embrace such an apparently great advantage. The two gentlemen posted off at once for P——, and obtained the desired interview with the lord of the manor. He was completely deceived by the gentlemanlike, off-hand manner, of the London broker, whose appearance, being handsomely dressed, and profusely decorated with jewellery, highly pleased him. Being accompanied by a late shareholder, he had no hesitation in promising him the grant of the sett, provided 50*l.* was paid down for the lease and expenses. This was at once done by the two gentlemen, who paid 25*l.* each. The plans of the extensive estate and mine produced, samples of ore raised shown, glasses of wine discussed, as well as things in general, when the landlord asked, "Will you not go down to the mine and see for yourself, Mr. R——?" "No, I do not want to see the mine; I've got a stone of one or two and the maps, that's enough for me. I don't care a d—n about seeing the mine; I'll get a company to work it, and that's enough for me and for you." To the landholder's surprise, they left the estate immediately after.

The "bal seller" started next day for London, after agreeing with his friend at T——, that whatever profit was made should be equally divided. Arrived in town, the "bal seller" at once went to work. Being known to be from, and largely connected with, Cornwall, he had little difficulty in obtaining credence from some gentlemen, to whom, as an especial favour, he displayed the glittering treasures. What could be more satisfactory? There were the ores, the maps of the estate, the plans of the mine, the copy of the grant, sufficient reasons for its previous abandonment, and one-half of the mine taken up in Cornwall. Now, reader, mark! he stated one-half the mine, not the profits! Could anything be fairer, or more open-handed? Nothing. First one gentleman went in, then another, until all went on as merry as a marriage bell. The broker (again now) had pocketed 1000l. for half the mine, set up his carriage, got on the Stock Exchange, and was considered a rising man. A magnificent house, handsome equipage, and swaggering gait, were his; well-furnished offices, in a good situation, were his also. Hearing of all this in Cornwall, his T—— friend came to town to see how matters stood. He was received and politely treated with unbounded hospitality, introduced on 'Change, and shown that but half the mine was disposed of. After spending 20l. or 30l. with his dear friend the stock-broker, in seeing the lions, &c., he returned to T——, perfectly satisfied that he had at last found a really independent, honest fellow. Soon afterwards, coming to town, he found his quondam friend in the Queen's Bench, and was sued for 1000l., as his half of the mine, being entered for that share in the pretended cost-book. The only way of getting out of the mess was by bankruptcy, which he endured, and was thus ruined.

Not a shilling was ever laid out on the mine. The period of our description runs over eighteen months. The landlord was quiet for six months, during which time the grant extended, by continued assurances from the broker and his friend, that a capital company was being formed, who would develop the mine fairly and fully. At length he, too, became tired out with promises, and desired to know when work was about to be commenced. His letters became more urgent and imperative, when a reply to one of his letters stated he might go to the d—l if he liked, as the thing was all one to the

broker. He could get a "bal" anywhere, and call it E—— S——, and it would be all the same to him. Disgusted with such a knave, he revoked the sett; yet the broker went on selling shares, until we find him in prison, as described.

One step in crime is sure to be followed up in quick succession by another. We next see our hero (for as one murder makes a villain, and thousands a hero, our criminal is fully entitled to the distinction) at Liverpool, playing the same game over again, but with rather diminished success, still with sufficient adroitness to be in the possession of considerable property, and lo! to marry a wife with very good connections and large property. This, of course, gave him an extended field of action, which he failed not to use with good effect. His wife's property was partially setted on herself, but what was not soon went off in extravagance and knavery; not a shilling of all the large sums he received was ever laid out in the mines he pretended to sell; yet losses like these are set down as incurred by ruinous mining—*pro phudor*. Like means produce like effects; he was again imprisoned, but after a period of some months was again released. We now lost sight of this worthy for some years, as he was hibernating on the plunder he had secured, and on his wife's narrow jointure. We next see him in public at Hull, again in full feather—a fine suite of offices, good house, wine and boon companions. Attractive advertisements brought plenty of business, and had he been anything like honest he might have done well as a broker; but "bal" selling was his evil genius, and that attended him like his shadow on a sunny day. He had too great a temptation put in his way by success; he tried a "bal" scheme, and got plenty of money, but alas! none went to the mine. One fine morning he was missed, and the "bal" found to have no existence. We now lose sight of him again for three or four years, when we find him figuring as an insolvent in Lancashire. The places of residence enumerated in his schedule were above forty, plainly showing by the brevity of time at each that there was no resting-place for him. Being remanded for two years, we presume he is still confined in —— Goal—a fit reward for such a villain. It has been calculated he had received upwards of 4000*l.* from his "bal" selling speculations alone!

One of his dupes in the E—— S——, hearing the mine was being reworked subsequently as the P—— Mine, took a journey from London to the spot, and, finding it was actually the case, waited on the landlord, and stated he had paid the "bal seller" 500*l.* himself, for which he only received a worthless scrip and one piece of ore,

about the size of a walnut. Though so severely robbed, he laughed
at his own credulity, but did not condemn mining, and again embarked, with great success—as great, he has told us, as he could
wish. His own words on the subject are—"I was a fool, but experience teaches me it was the best 500*l.* I ever laid out. I now see ere I
buy: I lay out 10*l.* to examine, before I invest 500*l.* By doing so, I
have gained many 500*l.*, and laid the foundation to gain many
more." We hope "Fides" is satisfied with our portrait. As he is
critical as to colour, we hope the "black" and "*done* brown" will
please his taste; if it come not up to his idea of a "bal seller," we will
try again. As we have before said, we waited our opportunity, and
can warrant the truth of the picture. We write this page in the very
room in which the E—— S—— transaction took place, having come
hither that we might have the exact particulars. The rest we know
from experience.

Many will ask, for what purpose are such pictures presented?
We answer, they are necessary as a set-off, to contrast the good, and
to warn unwary purchasers against "brokering bal sellers," whose
only qualifications are unbounded impudence, ignorance, and
knavery; to see for themselves what they purchase, and to know the
character of the men to whom they confide their money; to see that
it is laid out in the development of the mine, and not to trust to
flattering letters, that all going on satisfactorily; or to purchase bits of
worthless scrip or "slocking stones;" not to be taken by a well-
dressed exterior, splendid equipage, or profuse hospitality; to value
man, not appearances; and learn to judge the difference between a
worthy mine captain and a knavish "bal seller" and we shall hear far
fewer complaints against Cornish mining.

### The Mining Borough (St. Ives)

MISREPRESENTED AS CORNISH MINING HAS LONG BEEN OUT
of the "House," persons would have supposed that Cornwall would
have had ample opportunities for representing its interests in the
Senate, seeing that it returned no less than 42 members to the British
Parliament. A few reminiscences will, however, dispel the idea, and
will serve to show how mining was rendered subservient to the
would-be and were Members, and not they to mines or mining.
We can scarcely select a more striking example than the subject of

our paper, as it illustrates the difference in the two eras; and the contrast, we trust, will not be unfavourable to the present time.

St. Ives then was one of the Cornish boroughs distinguished by the euphonious title of "rotten," and was looked upon by certain worthies as a pocket borough, though the contending parties were so nearly balanced by their property rights that vigorous measures were necessary to secure the seats. The electors looked on an election as their "vested right"—in fact, their birthright, and as much a source of regular income as their trade in pilchards and fisheries.

Many well-known names are associated with this famous borough. Sir C. Hawkins, who worked the Great Shepherd's Silver Mine, near Michell, was one of its representatives. The Marquis of Normanby (afterwards Earl of Mulgrave), Mr. Tinley Long Pole Wellesley, Mr. Morrison (the London haberdasher), Mackworth Powel, and Mr. James Halse, did not think it beneath their dignity to become Members, and by such means as we describe. It is really amusing to peruse some of the grandiloquent addresses from them to a parcel of poor people, paying scot and lot, to whom the news of a dissolution of Parliament was the herald of a cheap drunken bout at the "open houses," and of a little cash in the pocket for their free and independent votes: the sum varying according to the number and wealth of the candidates, a strong opposition was always welcome.

In the immediate neighbourhood of the town are several mines, since proved to be very productive, but at this period untried for any but election purposes. As soon as there was a probability of a dissolution, a few loads of timber were sure to be brought on them, and every outward and visible sign of the mines being worked vigorously forthwith displayed. This would seem the more probable, as two or three gentlemen would as certainly be there also, to keep the game alive during the few election days, so as not to let the matter flag, by pointing out the vast importance which would be conferred on the town by having a good mine or two working so near it; but as soon as the election was over, and the seats secured, the gentlemen vanished, the timber was sold, and nothing more heard of the mines until the next Parliament. This ruse might have been supposed not to be successful more than once, but it was well known that Sir C. Hawkins was a bold mine adventurer, and was working the Great Shepherd's Mine entirely at his own expense. This inspired confidence, by which he was on several occasions victorious. He had a saying, when reminded of his promises, "Oh!

they are mere election promises, and go for nothing at all"—a principle he fully carried out here at least.

At length, Mr. James Halse, then a shrewd attorney, saw plainly how this might be turned to an advantage. He formed a company, who put the Wheal Trenwith to work with considerable success; and the advantages so often dwelt on and described followed as surely as a good mine goes to work in a locality. St. Ives became an active, bustling place—the people well employed, the market doubled, and its shops and trade greatly extended. Mr. Halse, as a benefactor to the town, its ships, and commerce, having always freights of coals and copper for those shipmasters who voted for him, was soon sent to Parliament, and retained his seat for many years, up to the time of his death.

One success followed another: more mines were opened, including Reeth, Balnoon, and others, until St. Ives district became celebrated for its good mines—a reputation it still maintains—St. Ives Consols, Providence Mines, and Wheal Margaret, being amongst the number affording good employment to a large population. At the passing of the Reform Bill, St. Ives was expected to be placed in Schedule A, and disenfranchised, the number of the population not equalling the standard. To remedy this evil, Mr. Halse purchased a moorland piece of ground in the midst of his mines, on which he erected a large number of comfortable cottages, which were immediately filled by his men, and the difficulty met.

Mr. Halse not only gained his reward by being chosen as a Member for the borough, but realised very large profits from energetically and properly working his mines. These until then worse than useless sources of wealth (seeing that they were rendered the agents of deception) may be seen in the cliffs near St. Ives, and are of a fine character, well worth a visit by strangers. Wheal Margery, the electioneering mine, now promises to pay well, and we sincerely hope the spirited adventurers will be rewarded for their money.

Nothing can exceed the moral debasement and low scenes enacted at these elections. Drunkenness, bribery, false swearing, parliamentary petitions, with all the concomitants usually attendant on such occasions: especially when the examples of turpitude and temptation were shown and offered by gentlemen of high standing, no wonder that poor, ignorant people fell, and were what they were. If the Reform Bill has done no other good, it has certainly remedied this. The people are far more elevated in the scale of society than formerly: now they have regular employment from

the mines, and are not so wholly dependent on the treacherous
fishing—they are more industrious, more comfortable, and less
inclined to the vicious habits for which they were notorious at
elections, presenting a favourable and great contrast to former
times, resulting, we opine, as much from the progress made in
mining as from other causes. We could multiply examples of
similar effects to a large extent, but we will confine our paper to this
once notorious, but now celebrated locality and borough.

May its example and effects be largely followed, for the advantage
of this and every other county where mining flourishes! It is a
necessary *sequatur* where these great sources of a nation's wealth are
properly and vigorously carried out. It has been proved so in Cornwall, even without the factious aid of its 42 members, or the sums
squandered on such occasions in venality and disgraceful riot, which
if properly applied would in many instances have been sufficient to
give a trial to mines, with such contrasts and effects as we have
endeavoured to show.


## THE KNAVISH CAPTAIN

IT IS FREQUENTLY THROWN INTO THE TEETH OF MINE
inspectors that they never give unfavourable reports—that he who
"peppers the highest pleases the palate the best." It may appear so to
some, but it is not the duty of the inspector to publish the document
he may draw out from his observation; it is only his province to
hand it to his employer, and let him do what he pleases with it.
Many a withering report is either consigned to the flames or for
ever hid. At the commencement of these papers, we portrayed
things and people as they are and were; then we were told we painted
everything *couleur de rose*. We are only doing our duty, and in pur-
suance of that we at length come to the bad as well as the good. In
describing them we shall not swerve from our intention or our path,
but go fearlessly on, turning neither to the right hand or to the left.
On this occasion we shall keep our incognito, though the characters
(for there are more than one) are well known in Cornwall, as well
as in London, where they have good reason to be.

One of these gentry, some few years since, visited "town,"
where his specious language, exemplary conduct, and pious demeanour soon won for him golden opinions and golden sovereigns

withal, wherewith he journeyed to Cornwall, and commenced a mine with a genuine Cornish name, which we shall call by the initial of C—— Mine.

Well, Wheal C—— went to work: tin was soon found; considerable quantities were raised and thrown over the burrow at a certain place as rubbish, though rich enough to pay—the reports from the mine continually stating that tin was always being discovered and raised. The fact of its being so was well known in the county; the shares advanced in price continually; the captain's friends bought largely into the mine. When the demand from Cornwall was known in "town" to be a "great fact," the "town" thought it the very nick of time to buy, but shares were not to be had. Tin was now returned, a considerable balance realised, shares rapidly advanced in value, a dividend was declared, and another at the next quarter's meeting promised by Capt. Jemmy; but at the next account, instead of a dividend, a call was necessary, and was made—not on Capt. Jemmy's friends, but on the London dupes, who had eagerly purchased shares in a "dividend mine"! Soon afterwards it was found that the engine was not of sufficient power, and not in the right place. Capt. Jemmy sent in his "resignation," to prevent being kicked out, the "bal" was stopped, and the materials sold. Capt. Jemmy attended the sale, and to the surprise of every person present bought rather largely of the tin-dressing materials at a cheap rate. When the halvans or refuse came to be sold, Capt. Jemmy was the purchaser. He well knew where the tin that ought to have been returned for the adventurers was thrown, and there he commenced active operations. Having the apparatus on the spot, and his family being brought up to the business, and working on the mine, they soon made handsome returns from that which he had previously represented would not pay for dressing. This source of wealth lasted for some years. It is said he realised many hundreds of pounds from this "speculation," as he called it—a piece of knavery worthy the most "cute Yankee" in that land of sharpers.

Time rolled on, the past history of the mine forgotten in "town," or lost in its maze of misfortunes and more gigantic swindles, and the name of Capt. Jemmy quite out of mind. That same fell destroyer exhausted the captain's halvans, and the captain (now a man of substance and sanctity) again revisited the City of London, armed with a well-written prospectus, as well as maps and plans of a mine; the name had been most judiciously altered, so as to prevent the shadow of a shade of suspicion. It was proved beyond controversy that the mine had sold considerable quantities of tin; the smelters' bills showed it, and all this from the mere refuse of the former workers. Affidavits of the men proved that they only left their work when they were actually driven out by the water—that the last month they worked they earned capital wages—that the engine was too small for the mine, and was in the wrong spot properly to develope the works. All this, backed as it was by a venerable personage who was known to be in the "connection," by his travelling certificate, as a preacher of the gospel, soon again procured him another company for the late C—— Mine. Immediate measures were strongly urged. There was a most capital second-hand engine, as good as new, to be sold, and only two or three miles off. Tin was high, and nothing could possibly interfere to mar their prospects: so far so good. A deputation was invited; a deputation came, saw the mine, took the miners' "depositions," viewed the capital engine and the captain's late works, which he open-heartedly offered them at what they cost him, though he had wonderfully improved them at a considerable expense. The "deputation" (who knew as much about what they saw as if they had studied a transit instrument instead of a steam-engine, or of the proper situation for a mine, as if they had been requested to point out the proper place for a new planet, or of the value of Captain Jemmy's plant, as if they had been asked to name the worth of the new picture in the National Gallery) returned to London, highly delighted with their visit, and with their captain, who they assured their fellow-directors was a pious, straightforward man, every way suitable to their purpose, and worthy of all confidence.

On this report orders were at once given to purchase the engine, and prosecute the mine with vigour, which was effectually done. A capital engine-house soon reared its white-washed presence in the proper spot, the captain enjoyed a handsome salary, his family found employment on the mine, the farce of throwing away halvans was enacted over again, some fault was found with the mine and its management, the shareholders in disgust stopped the "bal", the captain bought the refuse and the dressing apparatus he required, and has returned hundred of pounds profit by his second "specu-lation." The halvans are now very nearly done, and as the third attempt is said to be lucky it will doubtlessly be tried, unless the captain see this article, when he, being aware his motions and actions are known and pretty closely watched, may refrain, as he ought, from such malpractices. If the mine had been honestly conducted, there can scarcely be a doubt it would have been remunerative. Should it be again offered, we shall not fail to ascertain if Capt. Jemmy be in connection, when even his sanctity will not save him from exposure. The cloak will be stripped off, and the whited sepulchre (we beg pardon, we mean new engine-house in the proper place) shall be exposed. The author's name is well known to this worthy, and he is perfectly welcome to these remarks, and to this piece of advice—"Go thy way, and sin no more." Such, gentle reader, is a true Photograph. Have we redeemed our promise? Is not this a knavish captain?

We know another instance, in which a red-faced fellow, with stentorian lungs, and with a bouncing rough impudence that in London was mistaken for blunt honesty and peculiarity of manner, actually sold a mine in the City before he had seen the spot, or knew where he was to look about for one. At length he fixed on one; 1000*l*. were sent to him as the price to be paid for his discovery, as well as a goodly amount of "free shares," which he at once sold at a quarter of the specified price at which they were to be offered to the public. By these means he netted 1500*l*. in a few months, in addition to his salary, and was looked upon as a shrewd, clever fellow. At last the mask fell; the shareholders found the true character of their pretended friend, that the affected vulgarity was not the bluntness it appeared; that all these appearances were assumed, and that in reality he was a drunken, ignorant knave.

It is a melancholy task yet to draw other portraits of similar subjects, but, however unpalatable, justice must be done; we therefore proceed. A certain captain possessed a mine of untold wealth; the returns were something absolutely fabulous. At length, a noble lord was induced to join the speculation, and allow his name to appear. The company was to be divided into a large number of shares, so as to render the sad objection, a call, absolutely impossible. The wily captain modestly claimed one-fifth of the shares, and 1000*l*. for work done and the lease of the mine, for which he had paid the sum of 100*l*.! After the appearance of the lord's name in the list of directors, the shares, which had been previously but little in demand, were rapidly applied for, when the wily captain suggested that, as they could not possibly ever expend their large capital, it would be true wisdom to lock up the quarter part of the shares until they rose to a handsome premium, which they positively must do as soon as they were in a position to make their enormous profits evident to the public. An announcement accordingly appeared in sundry papers, stating that no more shares in the mine would be issued, a sufficient sum having been placed in the hands of their bankers to meet all possible contingencies. The bait took, the place was besieged by applicants, the shares soon went from 1*l.*, and were sold at 3*l.* to 3*l.* 5s. each. The captain, since none could be had out of the office, "obligingly" spared a few of his free shares at these quotations, to his most particular friends, however, giving his former customers the preference. By these means he sold all his shares at a premium, and left the country. A few months showed the whole matter to be a rank swindle, meant "but to betray." Many lost their all in this bare-faced robbery, and can painfully testify to the facts of the case, as it is and ever will be alive to their memories, the splendidly engraved scrip being the only thing they have to show in lieu of their banker's cheques, or their hard cash.

Another picture, and we have done. In this case, also, the appearance of extreme sanctity, good address, grey hairs, and a pair of green spectacles, did wonders. This rare performer affected to have become too old for underground toil, and was, therefore, about to undertake the still more difficult but less laborious task of "mine manager," which would afford him a much better opportunity of seeing things were properly managed. With this view he was determined to commence, *de novo*, a concern he had for years had his eye upon, but which his late engagements had obliged him to defer entering on; he had, however, relinquished all in favour of his favourite scheme (the fact was, he had been discharged for dishonesty), which he felt confident would afford provision for his declining years. He obtained the grant of a lease of an old mine sett, and, in company with a confederate, actually played off the horse trick. The silly dupe who purchased the sett found himself minus the sum of 500*l.*, and in possession of a most splendid (on paper) mining sett, with a name a skilful Welshman would be puzzled to decipher, much less to pronounce. The final settlement and handing over the cash occupied a week or two, an arrangement no doubt agreed on by the two knaves. When the second was called upon to take the property off the dupe's hands, his rejoinder was, "Oh, you've been so long about it I've laid my money out in another spec. I haven't got five hundred pence, much less pounds. You've got a valuable property, so be quiet. I'll help you sell it a profit, if I can." On appealing to the other, his reply was, "You applied to me, not I to you. I know nothing of your arrangements. The property is genuine and good, if you choose to work it; I haven't the capital to do it, and I told you so. You've got the property, and I the money, so both ought to be contented; I am. You may go to law as hard as you like: I defy you."

Our task is now complete, and having shown up so many in their true colours, we may perhaps be allowed again to return to subjects of a more agreeable and useful character. When called upon, we felt it our duty to give the shadows as well as the lights of Cornish life amongst miners. Where there are so many it would indeed be strange if all were virtuous: it would be against a natural law, and we do not claim that even for Cornish character, much as we rejoice in its genuine purity and magnificence.

## The Mining Village

A WEEK AT WENDRON, THE CENTRE OF THE TIN MINING district, can be well spent in observing the peculiarities of Cornish customs and character, which are to be found here in greater perfection than any other district in Cornwall. Here the old customs are still kept up as of yore, many words obsolete elsewhere are to be met with, and tin streaming and mining seen to perfection.

This parish is very extensive, but lying out of the high road from any of the principal towns, is not so well known as many other parts. Tin streaming was formerly, and at a very remote period, extensively practised in this parish, as well as surface mining, some parts of the surface for many miles being literally turned topsy-turvy, presenting nothing but a series of banks and pits, many of which are in dangerous proximity to the high roads and pathways. These excavations, of which no record as to the names of the workers remain, are attributed to the "old men," a term often met with in mining reports, and of which this is the true definition. In no part of Cornwall have the old men been more industrious, or more successful, than in this parish, the aggregate value of their tin on which dues have been paid, as proved by the books of the Duchy of Cornwall and the lords proprietors, exceeding the enormous sum of 3,000,000l. sterling, and it is more than probable one-third more might be added for tin on which no duty was paid, for there is smuggling in this as in all other duty-paying articles; we think we take a fair computation at that estimate. At present tin streaming is but little practised, the streams being exhausted; but the tin found therein is exceedingly rich. The tin found in all the tin mines of this parish is of very excellent quality for metal, it being remarkably white and silvery in colour. The stratum is granite, and the lodes generally nearly east and west, having a north underlie. Several very rich mines have been opened from time to time, and many more have been abandoned. Here were the formerly rich tin mines of Wheal Trumpet and Caledra, which gave thousands on thousands of pounds profit. At present, the principal mines are Wheal Lovel, Wendron Consols, Porkellis United, Calvadnack, Tregonebris, and Wheal Ann, all of which, though not paying dividends, are producing mines, and looked on as promising concerns. Besides these, there are many mines of lesser note, but in the gross employing many hands. These, with their necessary stamps, furnish occupation for many thousands of persons, of all ages, and both sexes. It is necessary thus to premise our photograph, that it may be the more clearly understood.

Our previous picture of "The Miner's Funeral" was enacted here to the very letter on Dec. 1, when no less than 3000 miners attended the obsequies of one of their number, who was killed at West Basset. "The Miner's Wedding," too, on Dec. 3, was exactly as described by our pen. Had we obtained the pictures from these subjects they could not have been more faithful. And now let us endeavour to transfer another, that will be recognised as equally true and equally frequent. The mine's descent has been given, and this task we had performed in the early part of the day of the poor miner's funeral. We found on enquiry underground why so few men were at their pitches—"Gone to berrin, Sir," so in the evening, when we returned, we sallied forth to take notes for our sketch book. We found the village swarming with miners, their sweethearts, and wives, some from the distance of 14 or 15 miles, come to the "berrin." After a great deal of trouble we obtained a seat in the bar of the village inn, which was crammed as full as it could possibly hold; here was the landlord, who is the very *beau ideal* of a Cornish miner in his true element, surrounded by his fellows, with a violoncello in hand, singing (he sings and plays with admirable good taste) hymns for hours, in full chorus. We have often alluded to the singular custom of singing psalms and hymns in tap-rooms of public-houses. Here, then, they were in full swing, showing the "straingers" what the Wendron (pronounced Gwendron) men could do; then Redruth would have its turn, and Illogan its trial, relieving the performances by discussing the merits of the deceased.

As soon as the first chorus was ended the landlord, an old Tresavean miner, stepped forward, and pressed me to take a tot (the term for a glass containing about one-sixth of a pint, saying, "Take hold and drink, strainger, you are as welcome as we." Having complied (it is deemed a marked insult not to do so), we were allowed to ask questions, which we did of a youth who had been explaining the manner of the poor fellow's death. On being requested, he furnished us with the following particulars:—"It was partly his own fault, Sir; he was 58 years old, and ought to have known better. I was in the pitch with en. Me and his son, who is my comrade, was working just above en, in the same pitch, not more than 14 or 18 feet off. The old man was shouling (shovelling) stuff away from under the stull. His comrade was breaking ore in the back. They had got a famous pile of ore broke, sure nuff, I dare say 5 tons upon the stull, they shouldn't have had half so much, when his comrade put in a gad, and a great big stone moved. His comrade called out, 'Stand out of the way, Uncle Jan, there's a big one coming; now, stand back, or you'll be killed.' 'Never mind,' said the old man, 'let en come,' but his comrade did not hit the gad again. His son and me called out to the old man to stand out of the way, when the old man went under the stull, instead of standing back; he called out he was all right and safe. As his comrade could not see him, he thought he was out of the way. The first blow on the gad brought down the great stone, and away went the stull upon the old man. I heard en squeak, and jumped, God knows how, from where I was to en. I wonder I wasn't killed myself. We were at en in two seconds, but he was smashed; he had the whole weight upon him. We got him out as soon as we could, but he was dead, and a wisht (melancholy) sight he was, but not half so bad as his son was, Sir: he was to be pitied, he took on so. 'Tis bad, Sir, to see a father killed."

After this narration, one of Wesley's hymns, appropriate to the subject, was sung, the landlord leading with the violoncello, and giving out the words. This finished, I enquired of the miner as to the character of the deceased; seeing so many persons present at his funeral, I supposed he was something remarkable. "He was a miner all his life, Sir: his name was John Moyle. He used to work at West Basset and many other large copper mines: he was a tributer, Sir, a very steady man, and saved up a good bit of money—besides, he was in three clubs. He was a good-natured man, and very much liked by everybody—almost everybody that knowed him were here to-day, and we are all sorry to lose en in such a wisht way. If he had died in his bed we shouldn't have cared; but to be killed and smashed up like that seems hard: but I suppose that will be my end; we are all of us liable to it at any minute, and we know it. I've seen many a bright fellow killed, but not smashed up like he was, but that can't be helped. It was partly his own fault, but his comrade blames himself as much for having too much on the stull. But he's dead and gone, and we hope his soul is gone to heaven." Then another hymn was sung, and another chapter on the abilities of the deceased as a miner. I noticed no allusion was made to any vices or peccadilloes of the dead comrade; it was really *de mortuis nil nisi bonum*—this would not be suffered. This is a good trait in their character, and should put fashionable scandal to the blush. At 9 p.m. the place was deserted.

On the following evening, after being underground the whole day, I repaired to the same spot, determined to find out where the innkeeper had acquired so much musical good taste, for he played and sang admirably, when I was surprised at the following colloquy: —"You're the gentleman that's been down inspecting the Consols, arn't you?" "Yes."— "Then I was at Tresavean when you was there. Do you know a man called Jenkin?" "Yes."—"He'll be here directly: he thought it was you. I do know you now well enough. Sit down. Well, I teached myself to play and sing too—we were all good singers when we got up to be men; and I endeavour to make good singers here now if I can. Boys, when they take it up young, like it; music makes them quiet when nothing else will; that instrument stops and prevents many a row, I can assure you, Sir; for, if they begin to quarrel, I takes out the article and strikes up a tune, when, if there's any noise, it is sure to be stopped by the company, and then I keep on until their tempers cool down a bit, and then the men are sensible enough. There's a deal of difference between miners now and what they were twenty years ago. Besides, when men are enjoying singing they don't get drunk. There's nothing got by a drunken man, he's a nuisance to everybody; and as no one will sit with them, they dispute what they've had, and there's no getting on; I've tried this plan on. All our young fellows that come here can sing, and I've no trouble; hardly ever such a thing as a row, and if there is one, it is amongst the old fellows, the young ones are right enough." Though I was here during the evenings of two pay-days, when upwards of 800 men were paid their month's wages, amounting, probably, to upwards of 3000*l*., yet I saw nothing like disorder or quarrelling; now, I attribute this, in some measure, to this singular method of attracting men's attention. The person I allude to is a man of great penetration, and was long employed as a common miner.

On the Sunday I visited the church of this large parish, ten miles long by not less than six miles wide, but of which vast area, and so densely populated, less than fifty persons were present to hear a very excellent discourse from the incumbent, who, I understand, is of most exemplary character and most charitable disposition. The chapels of the various sects of Methodism in the evening swarmed with immense congregations, when their singing propensities and abilities were brought into full requisition, and I recognised the tones and the visages of the men I had seen at the public-house of the village. The ministers in these conventicles were in one instance a travelling preacher, who gave a very homely but excellently adapted sermon; in the other, a mine captain officiated, with far more ability than I could possibly have expected. The congregations were orderly, attentive, and, apparently, devout, departing with decorum, and in the evening not a person, except a stray wanderer or two, was to be seen in the place where, the evening before, they had been practising the devotional singing of this their Sabbath.

To form a just estimate of the character of a people, it is absolutely necessary to see them in their social as well as public phases—to visit them at their homes as well as in their workplaces. This we did also, and found them humble, but content; their homes cleanly and neat, though the furniture in general is scanty and bad; the horrid cess-pool before the door, and the pipe universal in their homes as well as their workplaces. The people of this remote district, though poor and uneducated, are remarkably civil and communicative to strangers; they evince a readiness to give any information they possess, as to their locality, history, or employment, conveying a most favourable impression on strangers, which a few days' residence amongst them will assuredly confirm.

We conclude our notice thus briefly, as we purpose portraying other characters in other neighbourhoods, confining our notice of the mining character more particularly to the mining village.

## GREAT WHEAL VOR

*Labor omnia vincit* SHOULD BE THE MOTTO, AS IS THE PRACTICE, here. The term great is indeed applicable to this magnificent property.

To judge of its stupendous dimensions and capabilities a person should be something more than half a miner, or he cannot comprehend the subject. To persons not acquainted with mining details it would be worse than folly to expatiate, or attempt to declaim, on the tremendous undertaking we are about feebly to describe—an undertaking gigantic in every sense of the word, and in every part of its execution. No person but one possessed of confidence in his own abilities, and a knowledge of the extraordinary productiveness of the mine, would have dared to meddle with that which was deemed an impossibility—to fork the old mine. True, it has not yet been accomplished, but is hourly progressing to the consummation. Comparatively few fathoms have now to be overcome, and as the mine gets deeper the levels and works to be drained become less in extent, not presenting the formidable obstacles the explorers had to contend with in the upper levels.

To the inexperienced it may be necessary, indeed proper, to premise that the extent of the excavations east, west, north, and south, in winzes, shafts, levels, cross-cuts, and sinks, exceeds 30 miles! Only think of the Box or Bramhope Tunnels, and consider their cost—that neither of them exceed one-tenth the distance, and are comparatively at surface, not 200 or 300 fms. beneath the lowest attainable drainage—that this distinction makes all the difference. These tremendous (we speak advisedly in using this term) depths can only be attained by an amount of physical labour which must be tried to be appreciated. The agents find it no sinecure to climb 600 or 700 fms. every other day. To go to the bottom of one of the great mines, like the subject of our paper, requires not only nerve but strength to overcome the work. It is by no means an easy task to ascend the Monument, St. Paul's or York Minster, yet these are only one-third the height. These are also ascended by well-constructed staircases at a gentle inclination, not trodden on the stave of a nearly perpendicular ladder.

It would be, perhaps, better to place facts before the reader in the simplest form, and thus to lay out the preparations for commencing such an undertaking. First, then, order the carpenters to construct five or six miles of ladders, the ironfounder to cast upwards of 3000 ft. of pumps, none less than 14 in. in diameter, and some as large as 22 in.; the hundreds of tons of iron used in their construction may be more easily conceived than described or weighed. Notwithstanding the expense, there they are standing in the shafts, probably the highest and deepest monuments of man's ingenuity and perseverance to be found in the world. Iron columns, continuous for more than 1200 ft., to pump water from such depths requires an adjustment of machinery that no ordinary minds can arrange; to see the plunger-poles, to many the mysterious H-pieces, the complication of rods, clacks, door-pieces, and other underground appliances, confuses the imagination, whilst the ponderous compensation balance-bobs astonish it. Still these are only parts of the wonders of mining; the greatest wonder is how they are all paid for. Be it, therefore, remembered that this spot, not more than two miles long by one mile wide, has in the aggregate yielded not less than 3,000,000*l.* worth of tin; that a branch $\frac{3}{4}$ in. wide of good work, even if it be in hard rock, and at 50 fms. from surface, will pay; that at Wheal Vor there have been courses of ore worth 1210*l.* for 6 ft.; that where these things have been found they are and will be found again. Having heard so many persons speak and dilate so much on this celebrated place, we determined, as we were in the neighbourhood, to revisit a spot we had frequently trodden in our younger days. We first wended our way from Germoe Church town, through the Great Work Mine and Tregoning village to the Great Wheal Vor. Not a vestige of its former greatness met our view; the place was so altered that we knew it not. All was gone; even the old melting-house had disappeared—*sic transit gloria mundi.* That which we had in our youth deemed imperishable and permanent was departed—gone! We sat down on a burrow, and pondered on the mutability of mundane affairs. After a long pause, we came to the conclusion that "man cannot do without tin;" therefore, we took courage, as the Cornish mines must necessarily work. We then thoughtfully walked to an engine-house; seeing a notice in the window that no person could be admitted without authority, we, of course, acted on the warning, and got into conversation with one of the miners, from whom we gleaned a great deal of information; still he had that reserve so peculiar to the Cornish character of not letting you into the direct purpose you seek. A Cornishman always keeps a good back (if he be an able miner) on which to rely.

Having taken a lengthened survey, we were espied by some of the lynx-eyed captains, as they are called, when the following colloquy ensued. "Fine day, Sir."—"Yes, Sir; and a fine engine you have got here;" we were then looking at Trelawny's engine. "She is a fine engine. Do you know anything about engines?" On answering that we had a little experience in such matters, we were told "This is not our big engine, this is only our 85 in.; but our big engine is 100 in. My name is ——; what's yours?"— ——. — "Dear me; I shall be happy to show you over the mine; give me your hand, all our agents will be glad to see you." Under such able guidance we spent three hours in examining the surface of the mine, and we hesitate not to say, after examining nearly every mine in Great Britain, better laid out dressing-floors cannot be produced. Experience—practical experience—has been called into requisition; all the varied and multifarious adaptations of man's ingenuity for economising manual labour have been made available, and may be seen in operation. We visited the great steam stamps, driving 120 heads. Here we noticed an improved arrangement, deserving mention and compliment. All the heads were at work, rattling away in fine style, producing a noise that must be heard to be understood. The amazing extent of the dressing department had almost bewildered our imagination. After passing a good mile of continuous workings, and being fully convinced nothing in the shape of tin could escape the manifold and curious contrivances, we returned to the engines. We limit our description to the "big engine," perhaps one of the finest in the world; expressing our surprise at the exquisite finish, simplicity of construction, admirable working, and silence of the mighty machine, and complimenting, as in duty bound, the man who had charge of it for the good condition it was in, we waited again to ponder and think of Michael Loam, the constructor: if ability to construct such wonderfully great machines as this be possessed by man, what may he not do?

Our thoughts were almost naturally turned to the *Leviathan*, one of our wonders. Here we had a leviathan of another kind, indeed, working with all the ease and silence of a lady's watch. A cylinder of 100 in., a piston nearly the size of a man's body, with a stroke of 11 ft., equal beam, and lifting water from 1200 feet—that, too, a stream enough to drive a mill! We confess our gratification was extreme. After noticing the incipient man engine, for the ascent and descent of miners, we looked at two or three smaller engines, and were about taking leave of our courteous guide, when he insisted on our entering the count-house, where we found the manager, to whose politeness we are indebted for the following statistics:— That this celebrated mine is divided into 26,666 shares; that the whole expenditure from all sources, including returns, has been 277,136*l.*; that Wheal Vor, the part of the mine so called, has returned upwards of 27,300*l.*; Wheal Metal nearly 100,000*l.* These figures which startle the imagination, and almost create doubt in the mind; but when we have for facts that the mine paid profits at a time tin was selling at 35*l.* to 40*l.* per ton, and that it sold at times as much as 200 tons per month, as proved by books I saw on the mine, we were silenced, and could hardly complain of the present average price of 58*l.* per ton, although in a humour to quarrel with the smelters.

During our peregrinations we met with one of the last men who worked in the old mine: he assured us that the 274 fm. level contains the richest course of tin ever seen in Cornwall, and that it still exists there. He expatiated on its qualities, and showed how it could be made to pay for extraction; we were convinced of its practicability, as soon as that important part of the mine shall have been reached by the present company. This man's and our views we communicated to the agents, who coincided therein. We were then invited to join in their repast, dinner, not provided by the mine, every man bringing his own; there was a fried steak for one, fish for another, and broth for another, every one providing for himself to his own taste. After the repast we took a glass of grog, the captain's, too. We were shown some splendid stones of tin from a recent discovery, and some exquisitely executed plans of the pitwork, the dialling and sections of the mines, and the ground plan of the setts and works; in fact, everything that could be displayed were shown with a readiness and kindness that will never be forgotten. After a kindly shake by the hand of the older agent, whom we found to be an old townsman and acquaintance, we took leave of the surface of Wheal Vor, being determined at some day to go once again underground, and make a portrait of the extraordinary "bal," at this time causing so much talk in the mining world. We carefully examined every part at "grass", having heard so much of wanton extravagance being lavished heedlessly, and we do the agents and proprietors no more than justice in saying we saw nothing of the kind, not even the semblance of it; true, we saw magnificent machinery, but except a little justifiable pride in the possession and display of such (no loss to the proprietors), there is nothing on the mine to show any unwonted outlay. Indeed, after carefully pondering over the work done for the cost incurred, we wondered how so much had been accomplished for the money, and felt convinced, when this gigantic mine has had a fair trial, it will, by the aid of the appliances already provided, be found as good, if not a better mine than ever; but it must have that which all mines have—time, patience, and capital.

## THE "BAL MAIDEN"

NO SUBJECT IS OF MORE CONSEQUENCE TO THE SOCIAL welfare of society than the proper education and employment of the female portion. This then has been so often and so powerfully argued by philanthropists and political economists, that there appears little left on which to expatiate. Much, undoubtedly, has been accomplished, still a huge incubus has yet to be removed, a terrible evil has to be overcome. When Government enacted that no females should be employed in collieries, the prohibition should have extended to mines of tin and copper, though probably it was not aware of the evils which (at that period much worse than now) are prevalent in Cornwall. The indiscriminate association, in their employment, of the sexes naturally begets a want of modesty and delicacy, so important in the formation of character; whilst the masculine labour which females are frequently compelled to undertake, together with their being so long from home, render them wholly unfit to perform and attend to those domestic duties which should constitute the comfort and charm of every home, particularly that of the working man. Many a husband, many a well-intentioned man, has been driven to the pot-house from these causes, who had he had a comfortable hearth would prefer it to the haunts he now frequents, to his own and his family's ruin.

We by no means wish to imply that the evil is not gradually on the decline—we know it is so; our endeavour is to point out the evil as it is, so that by drawing attention to it, we may possibly assist in its more speedy removal, or entire suppression. It is fearful to learn the amount of demoralisation from this source. A few particulars of the employments to which these "bal maidens" are accustomed will show whether we complain justly or not.

We were on a mine a few days since at which the sampler was weighing off the ores sold at the previous ticketing; there being a scarcity of hands (men) on the floors, Mary and Nanny were called on to assist, on which two Amazons rushed forward to the work. Now, to those unaccustomed to such scenes, it will be necessary to observe that the ton of copper ore consists of 21 cwts., which are weighed seven times in hand barrows, containing 3 cwts. These girls were employed to fill these barrows with copper ore with the long-handed Cornish shovel, and carry them a distance of several yards to the scales; after weighing, they have to carry them several yards further to another heap, where they upset the ore, and then go to the pile to refill. This may appear very easy, and all very well; but let any strong man try the experiment, and he will find the task a most laborious one, and such as but very few can stand for two or three hours. These poor girls wrought most vigorously, being spurred on to the task by the challenges of the men; such as "Bravo, Mary!" "Well done, Nanny!" and amid coarse jokes and jeers that were not fit for ears polite, and certainly such as young girls had better not have heard. The sampler urged that they should get men to the work—that it was not fit employ for girls, and remonstrated on the subject, but to no purpose; the poor girls were obliged to execute the toilsome work, the perspiration streaming from their faces in copious floods. The lifting such weights from the ground is hard work; then the depositing it on the scale, again lifting, and the twisting to upset such a burthern, is quite improper to the female frame, and ought never to have been allowed. But so it is, and that too frequently.

On many mines, particularly tin mines, large numbers of boys and girls are employed in dressing the ores of that metal, which are always greatly mixed with other minerals and gangue. None of the ores are brought to market until it is in a state termed black tin—that is, deprived of as much foreign substances as possible by the action of fire and water, a most tedious and costly process. In many of the larger mines scores of "bal maidens," as they are termed, are employed, and in certain parts of the work an almost equal number of boys. Whilst at their work, and under the supervision of the master dresser, all goes on tolerably well, save coarse joking, and that continual association we so much complain of; at meal times, and on going to and from their work, it is almost impossible to prevent that conversation and rude behaviour we so much wish to see prevented. On some of the larger and better regulated mines there are separate and comfortable dining-rooms for each sex provided, where decency and order are strictly observed. This is a step in the right direction, and a great improvement. The evil is known to, and admitted by, all mine agents, who regret the necessity, but can see no remedy for it. The same supineness formerly existed with respect to females employed in collieries; the agents could not possibly see how the evil was to be avoided, until the public saw for them, and compelled them to adopt an improved regime.

The "maidens" are usually sent to the mines at the early age of from six to seven years, where they are taught to assort the ores, after which they learn to buck and jig them—that is, bruise and separate by water the ores of copper, lead, and zinc; a most laborious species of work, particularly the bruising of the ore, which is done by striking the pieces with a heavy hammer on a flat piece of iron. In the larger ore-producing mines, huge crushers, or rollers, are employed; in the smaller the "bal maidens" have this work to do; but, if actively pursued, it is too hard work for females. In tin mines these girls attend to the frames; that is comparatively light and clean work, but the constant exposure to wet is unfit for females; whilst the continual association with the men and boys, as we have before said, is highly improper. These poor girls remain from the early age we have mentioned, until they either get married off to some of the miners, or die of consumption, which carries off hundreds, annually.

The hard work is not the greatest calamity of which we complain, that is a mere physical evil; what we most deplore is, that when called to take upon themselves the duties of wife and mother, they are totally unfit for them. How can the moral standard of society amongst the lower order be raised by mothers and sisters with such education and examples? it is utterly hopeless. Taken from their hearths at so early an age, and kept at work for ten hours per day, they have little opportunity, and less inclination, to attend to the domestic and matronly duties so necessary to their future culture, and well-being. Their being associated in such numbers, and before men, a spirit of rivalry in dress (perhaps inherent in all women) is soon engendered, and every attention—all their thoughts and earnings —are devoted to this method of making themselves attractive. To see the "bal maidens" on a Sunday, when fully dressed, would astonish a stranger; whilst at their work the pendant earrings and showy bead necklaces excite the pity as well as the surprise of the thoughtful. All desire to save a few shillings for after-life is discarded, and nothing but display thought of. This is carried on to an incredible extent, and all the preaching in the world will never interfere with the wearing a fine bonnet or shawl, or an attempt to imitate the fashions of their superiors. Rivalry is the order of the day, and thus many are led into temptation. We see no present remedy for this evil, but trust machinery, and the non-employment of boys underground, may produce a mighty change, and oblige those who ought to be employed in domestic duties to be so engaged to their own honour, their husbands' comfort, and their children's blessing.

Machinery is rapidly effecting a change, and we hail every improvement in that department as a real blessing to miners; but where it dispenses with female labour in such situations, we rejoice in it as a grand and effective effort towards the domestic comfort of thousands, and as a help towards the elevation of the social position of mankind generally. If the Legislature be called upon to protect the poor little "bal boy" in his tender years, surely it is in duty bound to exert its powerful aid in favour of the "bal maiden;" and if the employment of girls in coal mines be improper, why is not such work improper in mineral mines? We know and feel we shall encounter a host of opponents, rich and poor, to any Government interference; that the cry will be "you will stop the mines," "vested rights," &c. And we know, too, the silent, subtle power of man's ingenuity—now rapidly developing itself—will effect that social good his cupidity and avarice would assuredly deny. We trust in Him "whose ways are not as our ways," and abide His time with confidence.

## ST. AUBYN'S DAY

THERE HAVE BEEN SO MANY DERIVATIONS ASSIGNED TO THE extraordinary term St. Aubyn, that to speculate as to the true one would be a task above our ability, and, we presume, not worth our trouble and research. Suffice it for us, if the sacrifices and libations offered in his name be indicative of his attributes, he was what is usually termed a jolly good fellow,—one of the "friars of orders gray," possibly, or of the brotherhood of Friar Tuck, and probably the latter: to use a vulgarism, the term means a jolly tuck out.

At mine account days, as we have before shown, where dividends are being declared, or where mines are really progressing toward that consummation, it is usual to have a good substantial dinner provided at the expense of the mine, at which the purser and manager preside, the company consisting of such adventurers as choose to attend, invited guests, as well as the captains and other agents of the mine. The proprietors then have the opportunity of questioning the captains as to the state and appearances of the mine, which can here be illustrated, and shown on the plans, maps, sections, &c., in a far more intelligible and lucid manner than the most elaborate report can possibly do; of examining the purser's accounts; ascertaining whether the wages paid in accordance with other places, or suggesting any improvements, or what they may deem desiderata. In such remote, out of the way places as the mines are generally situated, such an arrangement is absolutely necessary, if the account be held on the mine, which is generally the case where the shareholders are Cornish people. This custom has its advantages. Adventurers become acquainted with each other, and the benign influences of a good dinner, a glass of wine, and a handsome dividend, are not to be disregarded, leading, as they frequently do, to the ready payment of calls on another less fortunate speculation. These account days are usually well attended. Under London, or distant management, these meetings are impracticable, the shareholders being frequently distributed over wide distances, and at most but one captain in attendance, his presence involving as much expense as the count-house dinner.

Having previously tolerably well explained count-house doings, it would be tedious here to revive them. When the ceremonies of St. Aubyn's day comes to be explained, we fancy the extensive provision usually made for account day will easily be demonstrated. As a matter of course, all shareholders have an equal right to be present, and to invite a guest, which privilege is pretty generally acted upon (not to any injury to mining, by-the-bye, as many of shares are disposed of on such occasions), so that ample supplies may be required to be forthcoming.

All the cold viands, the cut and uncut joints, pies, pasties, &c., are carefully put by until the morrow, together with the half and whole bottles of wine (not many), and a modicum of gin or other spirit usually allowed by the adventurers, so that the captains and their friends may have a little jollification their own way. If there be not enough left to make a good spread, the captains usually subscribe a few shillings to make it up. It is usual at the account for the captain to offer to go underground with any shareholder or his friends on the next day, for their satisfaction. This compliment is frequently accepted, especially if there be strangers; it is the captain's spare day, when his attending a person underground does not interfere with his duties. They usually invite the captains of the neighbouring mines to the feast of St. Aubyn, which, as may be supposed, is seldom refused.

To see the Cornish mine captain in his true phase this is, perhaps, as good an opportunity as can be offered. Before such severe critics as the majority of adventurers generally are, and always pretend to be, the captains naturally feel a certain degree of reserve, and exercise a restraint upon their demeanour, conduct, and expressions; answering any questions of garrulous shareholders in ambiguous terms, and with quaker-like brevity and indirectness; as they are certain, if they ever so unintentionally err in judgment, or fail to fulfil the slightest promise, the opinions uttered will assuredly be brought against them at a future day; particularly if there be not an increased or good dividend. A degree of frigidity and formality of manner is thus engendered and displayed wholly unknown to them on the day of our titular saint, whose votaries delight in nothing so much as free and easy intercommunication of science and sentiment, which they invariably make the most of to his honour.

We lately had an opportunity of witnessing a fair average specimen of St. Aubyn's day. The adventurers, at their account on the day previous, had declared a handsome dividend; most of the largest shareholders and their friends were present,—a goodly company of about fifty gentlemen were there. Being an invited guest, we, of course, only entered the room when the business was over. The captains were busily explaining the maps and plans, as we have before said, when the chief captain asked if any gentlemen present would like to go underground on the morrow. The invitation was accepted by two or three shareholders and gentlemen besides ourselves, being about to do so officially for one of them. Suitable habiliments were provided, and at 10 a.m. we all descended the mine, the party consisting of three of the mine agents, two shareholders, two strangers, and ourselves. We remained examining the mine until 4 p.m. On our ascending, changing our clothes, and entering the room, we found the captains and agents of several mines assembled, as well as some of the adventurers, who had come to see how the adventurers had fared in their, for them, difficult task; these, too, joined the party in their repast. Who, on a black heath, can resist the temptation of a good dinner, good companions, and good punch? We think there are but few, and those few certainly not connected with mining.

Instead of the purser presiding on this occasion, the chair was taken by the captain; but, oh! how changed the scene! Instead of the demure, cautiously spoken, retiring individual of yesterday, he was the very reverse; full of activity, spirits, joke, and conviviality, carving away as if he were catering for giants. His first intimation to the guests was, "Now, gentlemen, this is our day, and I hope you will all do the best you can to make yourselves comfortable, and enjoy yourselves your own way." After rallying the underground tourists on their great exertions, he turned to the strange captains, and without giving time to answer a single question, asked a thousand and one of them as to their mines; and then triumphantly turning to one of the adventurers—"Never mind, there is not many will beat the P—— Consols yet. Will there, Mr. S——?" This was responded to in chorus by all the captains present, and cheerfully acceded to by the rest of the company.

The punch went merrily round as long as the tools lasted, when sundry drops of gin toddy were indulged in. As a matter of course, healths were proposed and drunk; success to P—— Consols Mines, and all the usual routine, applauded to the echo. We observed, too, that the coldness of bearing on the part of certain shareholders was considerably modified by a descent into and examination of their own property. They were far better satisfied when they saw what the so-called enormous sums had been expended for, and were far better satisfied of their agent's skill and abilities than they had previously been; and were equally astonished to find on the St. Aubyn's day the frigid, guarded captains were really communicative, intelligent, jolly fellows, as wholly different as possible from the agents of account day.

The life and spirit of the company on all hands increased in warmth as the hours grew later, when the guests departed, after securing a few specimens of their own breaking as mementos of their valiant undertaking, most of them insisting on the captains that could be spared jumping into their conveyances and accompanying them to the nearest hotel, there to finish the St. Aubyn's day. Thus it is under the present *regime;* formerly, it would have been a matter of difficulty, if not impossibility, for either party to have executed the latter part of the business, the libations in the count-house rendering it out of the question. We have only to do with what was done on the mine. We seek not to lift the veil from what transpired at the hotel, but we may say we heard (but that, fortunately, is no evidence) that the small hours chimed ere the company could be persuaded to part from each other's society, so gracious had they grown, and thus ended St. Aubyn's day.

We would advise every shareholder to act upon the principle enunciated in this paper—to visit the mine, not only on the account days, to receive their dividends or enquire into their affairs, when all is formality, and strictly business transactions engross the whole of their time, when the agents act, as we have seen the reason for, under constraint and reserve, but accept their invitation to visit and descend the mine, and to dine with them on the St. Aubyn's day. You shall then see them in their true characters, and glean more real insight into your mines and their management in one brief day than can otherwise be obtained by many attendances at the stated and set time of account meetings.

When man feels he has no unnecessary restraint on his actions, when he is assured within himself his words will not be brought up in judgment against him for an accident over which he probably has no control, and when he feels his employers have seen and know the real situation of the mine, and in a measure identify themselves with the undertaking, he takes confidence, and at once unbosoms his thoughts to, and makes friends of, his adventurers. No better opportunity for this is afforded than on St. Aubyn's day.

## Gwennap Pit

A CELEBRATED STATESMAN AND WRITER HAS SAID, "TELL ME the amusements of a people, and I will tell you their peculiar characteristics." There is great shrewdness in this remark decidedly, for much may be gleaned of their idiosyncracy as developed in their social enjoyments. If the Cornish, as a people, are to be judged of by this standard they will have a good position awarded to them for humanity and kindness, to which they are undoubtedly entitled; still it must be confessed that a great deal of the old leaven of superstition is to be found lurking in what would appear at first to be refined and strictly religious observances.

There is so striking a resemblance not only in language but in habits and customs of the Cornish and Welsh people as would indicate they were originally descendants of the same stock; both pride themselves as being the legitimate representative type of the ancient Britons; how this may be we know not, certain is it, however, that formerly both were servile victims of superstition. Though fairies, piskies, devils, ghosts, death-bed tokens, signs, and other popular fallacies still have their votaries, yet at the present day they are at a far greater discount in the dukedom than in the principality. From the towns these preposterous follies are nearly banished, as well as from amongst the more intelligent of the mining village population. Education, and the improved style of modern pulpit declamation, has undoubtedly contributed much to this desirable end in our western districts. Under the former dreadful influences the Mormons have probably made more converts to their monstrous and blasphemous system than amongst any section of people in these kingdoms. By the same aids Johanna Southcote was successful amongst the most ignorant classes of the benighted and barbarous parts of Yorkshire and Wales; but in the county of Cornwall neither of these "churches" (heaven save the mark) ever took root or had an abiding place.

John Wesley was the man of and for his day. Born, as it were, for the period, he with profound wisdom directed and made use of the popular prejudices, and sought thereby to convert error into truth. Using the force of self-reasoning on their own premises, he encouraged their belief in things unseen, unknown, and unfelt by many; he taught that there were moments of pure happiness enjoyed in this world by the virtuous character the vicious could never attain; and exhorted them to repentance by the call of the divinity implanted in every man's breast, endeavouring to inculcate Pope's beautiful verse—

"What conscience dictates to be done,
    Or warns me not to do,
This teach me more than hell to shun,
    That more than heaven pursue."

He, with fervour and assiduity, as remarkable for their power as for their rarity, pointed to the dictates of that "still small voice," and portrayed to the until then neglected population, with an earnestness and zeal hitherto unknown, the fearful consequences of the Divine wrath, and the folly and danger of wrestling with God by resisting his impulses; for the subjects of his discourses taking the most impressive texts of scripture. At first, this minister of good was treated with ridicule, indignity, and persecution; but a mighty change was at hand; the good seed fell into good ground, a few converts to the new teaching were made; these became consistent members of society, instead of the reckless, rough characters they had previously been. Of these some were selected by Wesley to further the important work he had undertaken; this was a great and politic step towards the rapid progress soon afterwards made. Chapels sprung up on every side; the simplicity of prayer, without the monotony and repetitions of our otherwise beautiful Church service, the constant change of ministers, the adopting more convenient hours of worship, the shortening the services, the change to an entirely new system of psalmody, in which all could enter, the fervour and inspiration of extemporaneous prayer and preaching, in contradistinction to the apathy of the clergy; and, in some cases, the novelty, in others the fashion, drew vast congregations, amongst whom were many

"Fools who came to scoff remained to pray."

To such assemblages as these, and from such causes, the celebrated meeting at Gwennap Pit owes its origin. On great occasions, such as holidays and festivals, the crowds were so great that no rooms or buildings were sufficiently capacious to contain a tithe of the concourse. Hence the necessity of out-door services, of which the subject of our paper is the most remarkable and most popular.

In bringing his congregations together on such occasions, Mr. Wesley had a happy method of selecting situations as impressive as possible from some grand natural peculiarity, to which, in his eloquent discourses, he could effectively refer for evidences of Almighty power and design, and which, under such circumstances, told with astonishing effect on these sons and daughters of toil, so little accustomed to practical illustration. Near St. Michael's Mount is a large rock, called the Chapel Rock, or, sometimes, Wesley Rock. On the sea shore, or sandy beach at its foot, is the spot on which thousands have listened to his preaching from the eminence on such topics as the miraculous draught. His audience were, two-thirds of them, perhaps, poor fishermen and tinners. With what thrilling effect would—"I will make you fishers of men" fall on their ears in so appropriate a place! Whilst preaching in the midst of the sublime but wild scenery at the Land's End, standing on the hill immediately behind the then extreme and tapering point, he is said to have extemporised the simple and beautiful hymn commencing—

"Lo! on this narrow spot I stand,
Betwixt two roaring seas."

Lanyon Quoit (a druidical altar in the parish of Madron, in the neighbourhood of which are many stone circles, said to be temples of these ancient priests) was also a favourite resort. Here, in the wilderness, did this pioneer of the word of truth, with striking success, contrast with power their idolatry and licentious practices with the pure and holy doctrine of which he was the herald! But at no place did then, or do now, so vast congregations assemble, or is the original character of the scene so strictly preserved, as at the Gwennap Pit. Here now, as heretofore, on Whit Tuesday do thousands on thousands meet to hear the word of God preached forth under the glorious canopy of His heaven only.

Gwennap Pit is so called from the peculiarity of its form—a vast amphitheatre; probably, indeed, it is the remains of some old surface mining operations. The slopes all round, but in one place, consist of a series of steps, or rather seats, covered with sods, accommodating many thousands of hearers, by whom the preacher can be tolerably well understood when the services commence, as all is breathless attention; solemnity itself hushes all buzz or sound. As may be supposed, from the circumstances narrated, and from its being situated in the very heart of the great mining district of Cornwall, this is essentially the "miners' meeting." It is considered, and is in fact, one of their peculiar institutions, therefore should not be omitted in a description of the habits and characteristics of the mining population; this being *de facto* a mining service in every feature of the word. The preacher has abundant evidence on the spot on which to dwell; the wonders of His Almighty creation, as the miners in their daily work experience; and when dilating on the uncertainty of life, an appeal to the fact of the dangers of mining finds a response in hundreds of bosoms who have lost, as nearly all have, some friend or relative in the dangerous profession! No wonder, then, Gwennap Pit is so much held in reverence by miners, and the services looked forward to, and attended, with so much interest. Many days beforehand preparations are made by the surrounding neighbourhood to accommodate the crowds who visit St. Day (a very large mining village, near which the Pit is situated). Still, all cannot be accommodated: many, knowing this, bring life's substantials with them. From early morning, at which prayer services commence, until about ten o'clock, streams of pedestrians, equestrians, and vehicles of every sort, size, and description, pour in, as one unbroken tide, crowding every road and approach, each endeavouring to secure good situations. Within a radius of twenty miles foot passengers repair by hundreds; the choirs of the different chapels bringing their musical instruments to swell the chorus. As they walk they practice the hymns appointed for the services, this serving as a solace during their journey.

Custom has made this day more observed than any of the parish feasts or local institutions, to the entire exclusion of all the ancient Whitsuntide sports, save wrestling. This truly Cornish miners' game, year by year, is fast also falling into disuetude; the people evidently preferring the quieter and more refined sphere of life.

One great inducement to this change is the amazing number of young men who, from sobriety, correct demeanour, with a little self-respect and self-culture, aided by competent practical experience, have raised themselves to comparatively high situations as mine captains and agents, not only in their native county, but in nearly every mining part of the globe.

In Cuba, Mexico, Lima, Coquimbo, Jamaica, Australia, Isle of Man, Ireland, Yorkshire, and Wales, large numbers of managers and skilled labourers spring from this class of persons, who carry into those distant climes many of the practices and impressions received at such meetings as the "Pit." True, many sent out perish from that bane of the miner—drunkenness. This to them is far more fatal than the miasma of the swamp, or the blaze of the noonday tropical sun; yet even this is in some measure evidently yielding to the force of example and precept. This injurious custom is not carried to anything like such an extent by the agents and captains as it was only a few years since; their influence exerts itself more powerfully amongst the men than would appear to the eye of the casual observer, but to those who know their characters it is evident and palpable, and the total abstinence movement has a host of followers in this country. As may be supposed, a batch of this society's lecturers are to be found at St. Day on the Pit anniversary, urging with all the vehemence and rhetoric that they can command the grand advantages of temperance. Here, too, may be found the active representatives of the various sections of dissent from Methodism; but the "old connection" still does, and probably always will command a vast majority over the schismatics; old associations, and the memory of their founder, naturally lead to it. To witness so many thousands of these rough, hardy sons of toil and danger for that day resigning their work and their wonted amusements to assemble in the rude temple of the Pit, for the purpose of Divine worship (whether sincerely or not is not for us to say), is a grand sight, which cannot be contemplated without emotion and awe. At the commencement of the services, when the preacher gives out the hymn, the chorus of praise which ascends from so many well-trained voices pleases the ear, and thrills through the soul of the contemplative, raising its aspirations, and fitting it for the higher and more extatic duties of prayer and admonition.

The regular services are pretty much the same as those in their chapels, except that they extend over longer periods. The season of the year, in a climate so mild as that of Cornwall, renders this out-door preaching not only more practicable, but more pleasurable than being pent up in a large building; whilst the voice of the minister is better heard, not being lost or confused by echo or reverberation. The utmost solemnity and silence prevails; indeed, figuratively speaking, a pin might be heard drop.

Ministers of more than ordinary ability and celebrity usually officiate; the congregation consists principally of persons engaged in mining operations, with their families, in their best attire, who at an early hour of the evening retire, the younger branches to amuse themselves as they are wont, and as we have previously described. We say again, to see these people in all their phases, and to form a just appreciation of their character, it is necessary to accompany them in their amusements and devotional exercises, in public and in private, when we hesitate not to say the practises described in our paper will be admitted to be infinitely superior to those prevalent twenty or thirty years since, at which period badger-baiting, cock-fighting, wrestling, and such brutalising sports were in vogue.

We would advise those who have not done so to visit our Pit Day, for they can scarcely be said to have experienced the Cornish institutions without doing so: they will see for themselves, and rejoice that the people prefer the quiet, edifying, and certainly more rational and refining influences they obtain, by spending their holiday, as detailed in this paper, at the "Gwennap Pit."

## MINING AND UNDERMINING

THE NEXT TO UNIVERSAL COMPLAINT AGAINST MINING IN this and every other country is that the practice thereof is not generally pursued for legitimate purposes of the profession, or as mining should be: that these means bring it into disrepute, disfavour, and contempt, we are far from denying; indeed, from painful experience we know it to be so in many, sadly too many, instances; but we repudiate such practices as appertaining to "legitimate mining." Could but a tithe of the faults, trickeries, and mismanagement attributed to mining be fairly attached to it, we should for ever abjure the science, and blush for its professors; aye, and so blush that were we even a collier or a Cairne Kye man the glow of conscious guilt would be perceived beneath the grime on our cheek. We would not hesitate to advise all who call themselves miners to take up their traps and walk, to let merry England and her virtuous sons alone in their glory to work their own riches, and see if they, then, would find fault with each other—to let miners migrate to other more favourable scenes for their labours. But, as we know, the arrows of calumny fall harmlessly at the feet of truth, we remain in full confidence of their cause, and of its ultimate triumphant success over the foul stigma under which it has long laboured; and we trust, by our showing that undermining has long been represented and mistaken for mining, we shall be doing them and the public equal service—the former by doing them justice, and the latter by explaining the difference between the shadow and the substance, which at first sight it must be owned are so much alike as to be mistaken the one for the other.

Mining, as laid down in our English dictionaries, means the art and science of procuring metals and minerals from the bowels of the earth. Mining, as represented in common parlance, is too frequently understood as a system of barefaced robbery—a delusion, and a snare. Now, either the one or the other must be a fallacy. To illustrate our subject, we must draw from nature; the effects then (if correctly photographed) must be true, and we must see how they tell in our picture.

The subject of which we are now about to treat is—the fate of mining when unfortunately placed in improper and unprincipled hands, or is made use of for other than legitimate purposes. Our portrait will be known to many, as is our intention. A light set on a hill cannot be hid, and our mines shall not form an exception. Well, then, mining being the art of procuring metals and minerals from the bowels of the earth, who, pray, gentle reader, are the proper persons to be employed?—Propriety would answer, "Miners, of course!" Echo would reply, "Of course." In our instance, though miners long proclaimed that minerals and metals abounded, miners were not allowed to work the mine—that is to say, to manage and direct the operations. Parties wholly ignorant of the first principles, and totally incompetent to the task, took this, the main spring of success, on themselves; and, instead of applying the proceeds of such shares as were absolutely sold to the purposes of the mine, put the money into their own pockets, and declared that the shares disposed of were their own free shares, and not the company's; thus, out of thousands of pounds nominal capital, only a few scores were applied to develop the mine, and then in dribblets so spare that the miners, captains, and all were never paid regularly, but kept starving on from month to month, doing little or no work, being discouraged and disheartened ("No pay, no work" is the miners' maxim); until, at length, the strong arm of the law was called in, in shape of a sharp attorney, to compel payment (there are some of these gentry even in the mining districts, who fatten, too, in these apparently barren spots). This sharp practitioner was the "friend" of the poor miners in their distress: he kindly advanced them a trifle whenever they came to him with their complaints of not being paid by the purser, and requested him to get their money (which occurred at almost every pay-day). He, as considerately as kindly, at once served such shareholders as could pay with sundry slips of parchment, entailing a cost of two guineas each, by way of forcibly informing them of the fact of the poor men's deprivations. Though scarcely more than 100*l*. per year was spent on the mine, this worthy scion of the law boasted it was worth to him 200*l*. per annum during the whole period of its working under this company. As might have been expected, these proceedings came to an end; but what an end! —This skilled practitioner having an action (not for wages), issued process against two shareholders who he thought could pay his fees and costs at least. These he and they allowed to accumulate until they amounted to something frightful. When he thought they had gone as far as he deemed prudent, he insisted on immediate settlement. They, to protect themselves from incarceration, hurry off to the fountain-head, and placed themselves under the ægis of that imposing offspring of man's reason—Chancery.

We wonder if the ancient poet had the head of a Lord Chancellor in view when he idealised the head of a Medusa—whether he really had so fertile a brain as to figure his flowing wig curls as snakes, and the effect of these and his awful countenance as electrifying and paralising everything their influence fell upon, and turning them into stone? We must not thus digress in episodes, though the fact be patent to all in the stony heart of the law and its myrmidons: when once under the shield they, simple men that they were, thought themselves safe; but no, that would not do, for law is law, and can no more be carried on without money than can mining; so that somebody pay the piper, paper, parchment, pens, ink, and stationery will be provided. But there are sundry fees, refreshers, and other etceteras in Chancery quite as expensive as steam-engines, not half so useful, but far more powerful, as their expansion and power cannot be computed. To work a Chancery suit entails far heavier expenses than to work a mine. The result has proved the fact that about 300*l*. was expended to work the mine, which was then finished off by the law process; and about 3000*l*. has been expended to work the suit. The mine has since been sold for more than the actual outlay thereon; and the adventurers have been "sold" at not one-half their deserts, for their folly. The mine has been perseveringly and judiciously wrought by the purchaser, has been found to be what the miners predicted, and proved to actually be of immense value; but, *credat Judæus*! at the suggestion of one crotchety proprietor—and one only—this unfortunate, though valuable, discovery is condemned again to undergo the fatal glance, and to be again paralised by the horrid head.

These foolish adventurers, like thousands else, attribute their losses to mining. Now, candid reader, let me appeal to you—Do you call these losses chargeable to mining, or to mismanagement? We say, let reason rule. It has been proved beyond doubt that had mining or miners been attended to, or considered, the adventurers, in the first instance, would have reaped a rich prize, had they not entrusted their interests to unscrupulous, necessitous, and improper parties—had they followed the advice of those who understood the matter practically, and paid them regularly out of the funds which ought, in all fairness, to have been devoted to that end, and not have allowed themselves to be the victims of a sharp attorney; and had they not, in an evil hour, suffered themselves to have been ruined by one great false and fatal step. Yet, strange and mysterious as it may appear, with the sad example before them, the spell-bound party again rush with open eyes to feed and glut the insatiable maw of the monster, which antecedents should have taught them to avoid.

The picture is literally true, we grieve to say; it is the type of many: yet this is but a phase of the calumny under which mining so undeservedly labours. The injustice may yet be further illustrated in this very instance. There is not an individual connected with this sad affair but what will (if he go into the *Gazette* for the next 20 years) make mining the stalking horse and cause of his ruin. Though he has not expended 5*l*. in the adventure, he will never cease harping on the subject, and endeavouring to make the world believe he has been a martyr to the cause. The bitterness of his disappointment in allowing so great a prize to slip between his fingers is a source of deep and continual reflection: the wealth that was within an ace of being his haunts his mind. Had such parties but common sense they would, like many others, not only alter, but have good reason to alter, the burden of their tale; and, instead of gloomily condemning

the shadow, look by the light of reason at the substance. In place of
casting obloquy on mining and its professors, hold them up, as they
ought to be—the fountains of wealth, the sources of comfort, and
the mainsprings of our national wealth.

## THE MINE ADVENTURER

THE ADVENTURER! WHAT AN AWKWARD WORD TO BE USED;
but were the English language to be ransacked, not one could be
found more appropriate to, or significant of, the person or situation
intended to be represented by the noun substantive. So protean is
the character, that we confess the difficulty of the task we undertake
in its description under its various phases, and approach the duty
with diffidence. So very different, and so changeable under varying
circumstances is he, that we scarcely know in what point of view to
take him; so perfectly altered does he become, even under temporary influences, that we scarcely recognise him as the same individual. The proverbially sensitive chamelion is not half so un-
certain, for he merely changes the colour of his skin; the adventurer
wholly changes, not only his outward appearance and manner, but
his very nature and inward man become metamorphised, and
subject to "passions varied as the varying hour." We must, therefore,
proceed to secure the best portraits we can, leaving critics to select
which they please, as bearing the most correct resemblance to individuals within their ken, who are included under the denomination
of "Mine Adventurers."

To begin at the beginning, we will first portray the novice, or
young Adventurer, who, taken by the bait of a glowing prospectus,
the advertisement of a confidential and information-giving broker,
the blunt persuasive conversation of a wily, crafty old captain, or the
advice of a disinterested friend, becomes by one means or the other a
a member of their fraternity—for good or evil depends far more on
himself than is usually supposed. These parties are frequently persons
who have been most vociferous against speculations of every kind,
but more particularly mining. The practised bal seller delights to
hear young men with plenty of cash inveigh roundly against his
craft; he marks his victim, and after due precaution, preparation,
and time, is certain of his prize. By dint of repeated attempts and
applications, the novice is induced to take a small interest, and invest

a trifle he knows he can well spare; flattering accounts, with an advance in the price of his shares, completes what persuasion had failed to do. He enters on another scheme; this, too, perhaps enhances in value, and he effects a sale at a profit, most probably a fatal one for him, as he now supposes (this is the case nine times out of ten) that he is a judge of the value of such properties. If this "fond delusion" once gain possession of his soul, his fate is sealed. To render himself certain on this point, he visits the scene of his speculation, that he may become most thoroughly acquainted with its bearings—fatal resolve!

In this, the chrysalis state, an embryo Adventurer may be easily known by his sanguine temperament, the extreme satisfaction and anxiety with which he listens to any tale of mining success, however improbable—the doubt and derision with which he hears any facts of a contrary tendency—throwing all the blame of non-success on the shoulders of the unfortunate agent, and with admirable self-complacency satisfying his own mind that things would not have been so had they been under his supervision. He, too, may be known by his excessive liberality towards the object of his fond anticipations. It amuses the initiated to observe with what ecstatic enjoyment he expresses himself on the glorious visions conjured up in his vivid imagination, and to hear his description of beautiful castles of ærial structure—grand indeed for the time, but, oh, how ephemeral! He, poor man, works himself up to a pitch of expectation by transient success and overweening confidence that makes the reverse far more terrible and afflicting than reverses usually are. Such are, and always will be, the conditions of fresh mine Adventurers, except extreme caution be exercised, or experienced practitioners be consulted, and their advice adopted.

The young Adventurer, too, in his devotion to his hobby, is generally most anxious to be elected to some office in which his determined perseverance and example may be of value. With this end, and with a view of knowing the ins and outs of the affair, he usually gets, if possible, on the committee of management, or into the secretaryship, auditor, or some other post, in which position he imagines himself secure. Poor man! Let him read our Photographs of the "Captain" and "Purser," and then see! If there be a "Heautontimoreumenos" in the company, woe betide him, especially if that long-named fellow be a solicitor! Nothing can then avail him; neither good conduct, good accounts, or a good mine, can serve him (though if there be any salvation it is in the latter). We know an instance in which a fault-finder of this class, holding 55-10,000ths in a mine, filed a bill in Chancery for some imaginary faults; where the mine has been for two years, at an expense of thousands of pounds; and now the complainant desires to withdraw the suit, so that he be paid his bill of costs! These, and other similar turmoils, soon alter the spirit of the neophyte's dream—the stern realities of active life and responsibility burst upon his view. If he be well prepared for the change he need not fear or flinch. No man ever passed through life without vicissitudes; if individuals are not exempt, why should we expect immunity for companies? In the former case we fall back on ourselves for resources; in the latter we only rely on others' endeavours and integrity, a want of confidence in which frequently begets the downfall of the entire enterprise. To this fatal error the young Adventurer (if not in office) too often gives a ready and early co-operation to his own and his companions' ruin. If in the situation he so ardently coveted, his disgust and chagrin at the want of confidence shown towards him is extreme, and he, out of sheer disappointment, not with the adventure itself, but from one or other of such causes, throws up his interest with indignation, vowing he will never again be concerned or connected with anything like a speculation, and his old word against mining comes spitefully out of his mouth. No doubt he means what he says at the time; but age works wonders—it gives men experience and grey hairs whether they will or not—it should give them wisdom, but does not in every instance confer that boon: where it does, we see the change for the better, and hail it in after life with joy, as in the old Adventurer.

It is said "a woman's first love never forsakes her during life." The same may be said of a man who has once adventured in mining; whether he has been successful or not makes but little difference— the fact is the same. If he has been fortunate he has good reason to increase his interest; if not he, like all others, endeavours to regain what he has lost. Independently of all this, there is a bewitching uncertainty about mining that becomes positively irresistible to its votaries, particularly if they be non-professional. Be that as it may, when the heat of temper subsides—when the evil day is passed over and is forgotten—when the time that mining was abused and decried is a matter of history—when metals are in demand—when shares are in a rising market—when every one is making money, and all is couleur de rose—the old Adventurer, at an early stage, feels the warmth of the cacoethes revivify him, and he issues forth again into the vortex of his old habits, expatiates on his former luck and experience, for the especial delectation and behoof of the then neophytes, by whom he is looked upon as a prodigy and oracle, they, as in duty bound, follow his advice and example. At this period he makes hay while the sun shines, and does not sigh for power or place, with its arduous, onerous, ill-paid, thankless duties. No; "experience bought is better than experience taught:" he practises the old apothegm, and does not find even this kind of mining so bad after all; thus he closes his career as an old Adventurer. The poet says, "there is a tide in the affairs of men, which taken at the flood leads to fortune." Few things afford more practical illustrations of the truth of the remark than mining. We have known scores of Adventurers who at their outset and early career in life suffered all kinds of privations and trials, but who by perseverance, and by not being warped or troubled with temporary difficulties and trials, have triumphantly shown success to be the rule, and not the exception. These gentlemen generally begin where the timid leave off, and thus reap the profit of their outlay; they watch the turn of the tide, and float on with its powerful aid to the haven of fortune.

It is no wonder that we see so many phases of character as we witness in Adventurers, from the commencement to the conclusion, as we have just attempted to describe; from the bland, kind, cheerful, liberal novice through all the trials of committeeman, chairman, &c., in all its responsibilities and cares; to watch these virtues clouded, perverted, changed to acerbity and morosity of temper, such as render him unbearable to himself and everyone by whom he is surrounded, until he at last, as we have said, quits the affair in disgust; to watch the same man under his experience, quietly visiting the mine and, instead of fuming and working himself into a fury over matters he now understands, seeing the captains and agents do their duties, that the committee and chairman are constituted of persons qualified for their situations, and not mere aspiring, sanguine novices. He tenders them his advice, and sees their orders executed; thus enriching them as well as himself, doing more good than all the ardent young Adventurers or knowing fault-finders in the world put together, to the infinite delight and profit of the officers of the company, who prefer having a person to watch over their actions who really does understand and can appreciate their labours, than a supervisor who knows nothing, and, therefore, continually finds fault without reason.

And now, oh! Adventurers, if you be successful in your *debut* be not over sanguine; if unsuccessful for a time be not discouraged; be not alarmed at a groundless panic; but, above all, do not be led away by or encourage the advances of a "Heautontimoroumenos." If you be induced, by untoward circumstances, to desert the profession, remember to take the tide at the turn—to begin where the timid leave off, and you, too, shall gain that experience and success which will ultimately place you in the situation we earnestly desire to see you—viz., that of a successful, very old, Mine Adventurer.

## Miners' Superstitions

PERHAPS THE INHABITANTS, PARTICULARLY THE VULGAR, OF every country are more or less imbued with these impulses of the human mind. It is found to pervade savage life, where it is mistaken for religion; it is, therefore, not at all surprising that such veneration is accorded to the unhallowed rights and barbarous customs as we hear are practised by the primitive races of mankind. In civilised life they are characteristic either of a weak understanding or uneducated mind: as a rule, it may be set down that the more ignorant a people are the more are they devoted to superstition. Even where education is given, the oft-told, long-cherished tales, handed down from family to family, become so engrafted and naturalised as to be almost identified with their very nature, requiring the experience of years of instruction to wean the descendants, and to eradicate their belief in these fancies. Nor are they entirely confined to the uneducated classes; frequent instances occur in which superstition reigns predominant in spheres where we should have supposed errors so glaring would have been expunged or impossible. It is to this force of habit we may attribute the remnants of silly credulity still to be found amongst our otherwise shrewd, well informed, generally religious, and well conducted mining population. The legends of many a foolish, idle tale are still told with the greatest decorum by old folks as having been told them by their sires; they, as duty bound, asserting their entire concurrence in the persuasion of their truth, inspiring awe and dread into the younger portion of the community. Though the relaters do not attempt to account for the probability of their stories, they would consider it heresy to doubt them. It is gratifying, however, to find that this gross infatuation is gradually on the decline, and will, it is to be hoped and believed, ere long become finally extinct. The credence in the existence of witches and "wise men," we are sorry to confess, is far from annihilated: these rascally professors generally gain such an ascendency over their dupes, that they dare not reveal their complaints of extortion, or the knaves so dexterously manage their dealings as to keep out of the grasp of the law; yet when opportunity has offered the magistracy have certainly most properly punished such instances with condign disgrace. Belief in witchcraft is not the only hallucination to which the miner is subject; some of his are very peculiar. The most popular is the faith in certain mysterious personages, termed "pixies," or "small people." They seem to be a kind of masculine fairies, but, unlike the softer sex of these diminutive folk, they delight in nothing but mischief. We have heard otherwise sensible, clever men declare with earnestness, and with a warmth which evidenced conviction, that they have been deceived by this agency—that though they have been in their own paddocks, only a few yards from their homesteads, and at other times could not be mistaken, yet being "pixey laden" (spell-bound) they had been unable to find their way, until some one came to relieve them by breaking the "charm"—that during this unaccountable influence, though they could not see these extraordinary personages, yet they could distinctly hear their laughs and jokes at the sufferer's dilemma. The reader may well exclaim in wonder, Can it be possible that such superstition is extant at this day? We say, Yes: and to this hour in some districts. The fairies, too, have their share of devotees, but these are more amongst the agricultural than the mining portion of society, the rings of the midnight revellers being supposed to attest the realities of their presence. In the mining districts no herbage of sufficient luxuriance is to be found to tempt these fairy dancers to trip the light fantastic toe.

If a thousandth part of the fables could be authenticated, ghosts might be pronounced plentiful; there is scarcely a lone churchyard or abbey ruins where they have not been seen. White owls are not uncommon, and the goat-sucker, or night-jar, is frequently met with, uttering their dismal and discordant cries. These may in some measure account for ghostly appearances, which are firmly accredited by thousands. Some years since, a farmer's son, near Penzance, met a justly-deserved castigation whilst attempting a practical joke in this character. A miner returning from the town, being pot valiant, was determined to combat his ghostship, even if he were Satan himself.

On meeting with this customer, whom he verily believed to be Satan come for him (the clown being wrapped in a bullock's hide, with horns and tails), he resolved to fight hard rather than be taken to Pandemonium before his time; in doing which he so used his staff as nearly to kill the would-be thought supernatural personage, to the great joy of the neighbourhood where these pranks had been a source of continual annoyance and alarm. The fool was also very properly punished severely by the authorities, and the miner commended for his pluck.

Amongst the imaginary influences affecting miners more immediately, may be mentioned dowsing, or a belief in the divining-rod, and flames of fire issuing from metallic veins. These are somewhat akin to the fallacies foisted on and believed by the public in the late animal magnetism and table-turning delusions. These, like their compeers, have been exploded as childish by scientific and practical men, and the impossibility of inert matter acting as supposed by any inorganic power displayed; yet they are so strenuously advocated and defended as correct and true by others of great celebrity as practical miners, and able, well informed men. Many of the principal mines in the county are said to have been discovered by these agencies, though we have never yet met with a well authenticated case, they being generally what the son had heard father say what his father had seen!

We have often been tempted to try the efficacy of the dowsing stick: whether it arose from our organ of veneration not being sufficiently developed, or from any other cause, the rod would not act in our hands. We also witnessed the feat performed by one of the most celebrated dowsers in Cornwall, who boasts of having discovered scores of lodes by the stick. On this occasion he proclaimed with certainty a lode of some kind existed in a spot indicated by the obedient dowsing rod. The mining company who listened to the tale, and were weak enough to believe in its augury, as may be supposed, lost the capital they expended to prove whether the lode were there or not. If people of this standing believe such childish tales, the poor miners may well be excused.

Dreams, too, come in largely for their share of patronage, as may be supposed, by the female portion; for this the authority of the Bible is quoted, and, with people so much addicted to credulity, this argument is conclusive. It is really amusing to hear the tales of the wonderful discoveries which have been made during dreams, and to see with what implicit reliance the dictates of nightly visions are followed. Not long since we were pointed out a place by an old man, who, with all the eloquence he could command, assured us his father had dreamt there was gold in that place, and had actually commenced an adit for its discovery, which cost the poor old fellow great labour and money he could ill spare, of course, in vain. "Yet," added our informant, "it will be found by somebody." Strange as it may appear, it is no less true, this spot has been since proved to be a very rich and valuable copper mine. So much for the old man's dream; he made a mistake in the metal, and did not take the method of double *entendre* used by the Delphine oracle, or he might have been equally correct with that celebrated ancient authority.

All the foregoing fancies seem more or less common in all parts of England, as well as with the mining population of Cornwall, and not to be wholly confined to the ignorant. One of their own superstitions is peculiar to miners, that is universal amongst them—their objection to any person under any pretence whatever whistling underground, for fear of raising the spirits. Strange as it may appear to a novice, the objection is always made, and, of course, is always readily complied with, or the men make most decided signs of discontent and uneasiness. Reasoning with them on the subject is useless; they will not permit it by any person on any account. Only a few weeks since we were underground with a most intelligent and really very clever mine agent, when for a moment forgetting ourselves, we indulged in a stave, when we were at once asked if we had ever been underground before, and were real Cornish, and yet dare to whistle there! "Because, if you intend to do that, you may go over the mine by yourself; you do not get my company." We were at once silenced, and did not again indulge in the schoolboy's solace, lest we should pay too dear for our whistle.

These errors, however they may be regretted, are innocuous, compared to the practices of reputed fortune tellers and witchcraft. The arch villains who practise these arts, when once they obtain celebrity, drive a capital trade on the credulity of the population; the chief business being the patronage of the softer sex, who, as of yore, are as curious as Pandora, and display an equal anxiety. One of the most celebrated of these wizards was Johnny Hooper, whose name, though the fellow has been some years in his grave, is still spoken of as that of a wonderful man; many of his predictions and discoveries being still firmly believed by the peasantry. Johnny was a "character" in his day, and has consequently left a name. We warrant there is scarcely a person in Cornwall seven years of age who has not heard of Johnny Hooper.

Born in one of the most out-of-the-way places in the county, and as profoundly ignorant as it was possible a human being could be, yet withal possessed of that half-witted low cunning which is frequently and correctly expressed by the proverb—he is more R than F—this crafty knave contrived to eke out a tolerable existence without (to him the greatest dread of life) hard work. In our boyhood we remember him living in a little cottage on a wild moor in the parish of Ladock, near Michell. We were of a large party of young folk who out of joke visited Johnny, and stole his grinding-stone, challenging Johnny to name the delinquent, which he was wise enough to decline; thus, at all events, showing him in his true colours, that he was shrewd enough to evade an exposure by substituting a joke, and that he knew nothing of the affair, but that we were bent on mischief; so Johnny was hailed as a wiser man by us, at least.

We were anxious to ascertain an authentic account of this notorious character's antecedents and history, and to learn by what means he had acquired his celebrity and ascendency over his fellows. His fame was co-extensive with the boundary of the county; people from all parts of it resorted to him for advice. If anything were lost Johnny was applied to for the discovery of the delinquent; if any deemed themselves the victims of bad luck, or under the influence of witchcraft, he was the oracle consulted—with shame it must be owned, in too many instances by parties who ought to have known better. Undoubtedly, the fear of an appeal to this personage has frequently been the case of purloined property having been discovered, and the articles stolen being replaced or given up, not unfrequently by the agency of this fellow, who has been known to mulct double fees for compounding and settling robberies.

We found Johnny's history to be pretty nearly as follows:—He was one of those fellows, found in every parish, who have a natural taste for idleness; consequently, always out of work and in rags, ready to receive a penny to do anything not involving labour. At a Christmas party in the village, Johnny, then a youth, was arrayed as a gipsy woman (a character he was well adapted to represent), and having been placed in a dimly-lighted room, and being first well drilled by a young lady of the party as to what he was to say to one of the guests who consulted him (a young, handsome, and rich widow), the story conjured up surprised the widow, who was delighted beyond measure, and declared the gipsy to be gifted with superhuman powers, as she had told her all the circumstances of her past life, therefore she could not doubt her ability to name her future husband, as she had done, though she had never dared to aspire to such a thought herself. This essay brought the fortune teller 5s. After a drilling by another of the party, Johnny detailed a soft but tender tale to a young lady, describing her devouted admirer, and promising all sorts of fine things, for which he was further rewarded by a like sum, as well as 2s. as his fee. Johnny was cunning enough to see his way clear; a light had burst upon him; he started as a professional fortune-teller, with considerable success. To ratify his ability, it is said he stole a horse, which he tied to a tree in a wood (Arrallus Wood). The horse being missed, Johnny was applied to for advice; he demanded three hours for study, at the end of which time, in ambiguous terms, he announced the possibility of the horse being found in the situation named; search proved the oracle correct. The news spread like wildfire, and brought this knave into an extensive practice, which he enjoyed to the end of a very long life, fully serving his end and aim—a course of sheer idleness.

Unfortunately the belief in fortune-telling still exists to a certain degree, this is to be lamented, but strenuous endeavours are being adopted by all classes of educated persons and all sections of religious bodies to exterminate this foul blot, as well as the other super-stitious persuasions still extant. We firmly believe they will ere long be successful, they have excellent material on which to exercise their philanthrophic endeavours, no people being more tractable or more willing to learn any instruction intended for their real benefit, which we hail these efforts to be, and trust, in a few years, there will be found an entire absence amongst these people of "Miners' Super-stitions."

## The Drop in the Cup

HOW OFTEN DO GREAT RESULTS ACCRUE FROM SMALL beginnings. The fall of an apple led to Newton's grand discovery of the most subtle and powerful of Nature's laws. Unpromising as the title of our paper may appear, yet as we profess to be merely the narrator of facts connected with mining matters and mining people, we, as in duty bound, record such as we deem will be interesting

144 CORNWALL'S MINES AND MINERS

and instructive. Our papers, we have the satisfaction of knowing, are read by the younger portion of society. Should but one be led to reflect and act upon the moral of our tale, our end will be served, our trouble cancelled. We proceed to our tale, which a poet might richly adorn, a divine improve. But it becomes us only to hold the mirror up to Nature, to depict things as we find them, leaving the garnishing to abler pens.

We are aware that instances of a similar character are not wanting to illustrate extraordinary facts connected with human progress. As in this instance, they are always founded on some of the simple laws of nature. The sublime teachings of the Great Architect only require attention to render them profitable to him for whom they were intended to be enjoyed. Man is placed in a region of magnificent profusion, if he knew but the true value of the privileges by which he is surrounded. He is the realisation of the eastern fable of Aladdin, and withal has the "little lamp," which only requires a little rubbing to discover the true value of its hidden worth. We particularly commend our remarks to youth, in the sincere hope they may profit by our "o'er true tale."

In the United Mines, Gwennap, some of the deepest and most dangerous of the Cornish mines, there are (as we have said in one of our previous papers) "Bal Boys" employed. We, in that article, dwelt largely on their sufferings, and invoked (not in vain) sympathy with their woes. We were long aware of their ingenuity, and could have mentioned the improvement of self-acting gear, introduced into steam-engines for lifting arbors and opening valves, to save the trouble of doing so by hand, as an instance, but then we were not portraying their natural capabilities but their wrongs. At these extensive works, in the 100 fm. level (with the adit 636 feet from surface) is a place at which a capstan or windlass is worked, necessitating for their erection and working an extensive excavation. From this point as a centre diverge levels to the different parts of the mines. Here, where the men are not at work, with the ponderous machinery, all is gloom and silence, save the footfall of an occasionally passing miner with his dim candle. In one of these galleries, or levels, termed a cross-cut, and out of the usual route of even these few miners, were employed two boys, the elder ten years and the younger no more than seven years of age, to work an air-machine. Only one was in this spot at a time. Their duty here was to turn a handle, precisely like that of a grindstone, which rotated a fan with amazing rapidity, to convey air to the miners through a tube to the backs, or places where they raise the ore, in bad air. The slightest neglect or cessation from toil on the part of these poor lads would be immediately detected by the miners as a matter of course, and chastisement severe enough consequently follow. The monotony of this unbroken, continuous round in such a prison house, besides the labour entailed by continually exerting the muscles of the back and arms, may be conceived but cannot be described. No wonder, then, that these lads looked forward to the period of time for their ascent to sun and air. Boys will be boys—marbles and tops have their attractions. Many a weary hour did they pine their lives away, in hope and anxious desire. Solitary confinement and hard labour for eight hours are deemed punishment for criminals; what was this?

At length one of them desired to have a watch, that he might count the hours, and know when they passed. But, in vain hope! how was he out of 2s. per week to gain one? Necessity is the mother of invention. He, by the light of his candle, watched a drop of water falling from the roof at long but regular intervals. The idea struck him that if they were regular they might be made to indicate time. Procuring an old tin cup he had found in one of the levels, he placed it under the falling drop, and found that it always dropped into the cup. Here was a point gained. He borrowed a watch, by which he found it filled in a certain space of time. To regulate it so as to suit his purpose, he reduced the size of the cup, so as to fill it within the hour, and his experiment succeeded to his entire satisfaction and comparative happiness. "I felt the hours pass," said he, "and listened to the 'drop in the cup' as a companion and friend. I never afterwards felt that absolute isolation, that perfect oblivion, I had previously done. Besides, it taught me to think, to reason on natural laws, to pay strict attention to little things, with a view to their application to greater ends. From that time forth I began to study. I went to evening and Sunday schools; endeavoured to raise myself by study and care. I soon went underground, and worked as a tributer. My now settled habit of attention to little things stands me in good stead in this era of my life. I have been successful in an eminent degree. Soon afterwards I was created an agent. Here, too, I found the same habits had the same effects. I owe my success and present position in life mainly to this little incident in my boyhood. My solace and comfort made such an indelible impression on my mind that it will never be erased."

This bal boy is now one of the most eminent of Cornish captains, and has under his charge property to the value of hundreds of thousands of pounds. We gathered these particulars from his own lips. Though in the sear and yellow leaf, he still speaks of the occurrence with delight. It was as great an era in his life as the apple in that of Newton, and affords, we think, an excellent example of the fruits of attentive observation by boys of little things as leading to great results; for little did the poor untutored lad dream, whilst working in the United Mines, that he would become the manager of some five times as large, through having watched the "Drop in the Cup."

## THE PITMAN

IN ESTIMATING CHARACTERISTICS, CONDUCT, OR QUESTIONS, the *audi alteram partem* should be continually kept in view, or, to use a mining phrase, "Fair tawing is fair tawing." Our pen has frequently been employed in portraying the duties and practices of mine agents and others connected with those pursuits; it now becomes our duty to say something more than we have hitherto done of the men under their charge. In doing so, we hope our purpose will not be misunderstood, and that our observations will tend to benefit both agents and men, as well as satisfy certain adventurers. We have before remarked that in some mines, where mutual confidence reigns, the men, if once in the mine, look upon their situation as a permanent settlement, and would lay down their lives in defence or support of it or their agents. Their interests they identify as their own, considering themselves as part and parcel of the property, and always speaking of it as "our mine." There setting and pay-days are as described in the Photograph on that subject: it must, however, in all fairness be admitted that far different scenes occur where an over anxiety to serve the company, or an endeavour to appear so, for the purpose of serving private ends, stimulates the captain, and he curtails (as is too frequently the case) the men's wages to starvation point. There is then an end to peace, quietness, and comfort: distrust and jealousy usurp their place. Each endeavouring to overreach the other begets that animosity which leads to the worst results. The miner, knowing that at whatever price he may offer to work he will be reduced, asks much higher terms than he means to accept; the captain, knowing this, proposes accordingly. These paltry and foolish misunderstandings beget severe heart-burnings, mutual distrust, and disregard. They also display a want of method and regularity unworthy of business transactions. If the captains be not competent judges of the prices at which the men can live at a certain price of provisions, or what they should earn so as to place them on a level of comfort equal to other artizans, they are incompetent for their situations: this is pre-eminently one of their duties. They serve their own as well as the adventurers' interests best by allowing the men a fair living, and paying them regularly at appointed times. Surely the labourer is worthy of his hire, particularly in so dangerous and injurious an occupation as that of the Cornish miner. We know districts in which mine labourers, calling themselves miners (late farm labourers), are employed at a few shillings per month less than a regular born and bred miner—we mean one who can buddle, jig, &c., and take the place of a pitman; stack a plunger-pole and fix it, pack it, or a piston in the engine cylinder, without patent contrivances; in short, a complete miner. This mode of employment without distinction, and a culpable endeavour to depress wages, has led to many of our best miners leaving the country. Thousands are gone, never to return. The complaint is general and confessed. There is no lack of men it is true, but of miners there is a falling off, particularly tinners. A man can scarcely ever be brought to know tin in all its various ores except he be accustomed to it from boyhood; and its manipulation is difficult.

So many persons being placed on mining committees because of their position in life, large interest in the mines, skill in financial accounts, and so many people being appointed secretaries and pursers of mines who have really no practical knowledge whatever of the subject, has in some measure led to this serious error. They, seeing by the cost-sheet, that their new captain employs forty men on the mine at no more cost than thirty had previously amounted to, applaud his zeal and the interest he manifestly displays in the undertaking to the echo, and he is declared a very clever, pains-taking fellow. Poor men! they little know how they deceive themselves. They erroneously suppose they are effecting a considerable saving in labour cost, whereas twenty good miners would do more work than the whole forty; aye, and do it better too. The mistaking *men* for *miners* is one error: another is a practice in tributing, which frequently leads to disagreement, unless stringent bargains be made—viz., paying the miners at a less price than the ores sell at: this is more frequently the case with tin than with copper mining. This evil may be easily remedied by practical men, but is too often confided to those who are not, and to those who know not how to value the ore; hence the unpleasantness.

We know, from painful experience, that it is impossible to please all—to satisfy avaricious or spendthrift labourers, or to please parsimonious adventurers—committees consisting of tailors and retired costermongers (as is not unfrequently the case), who know nothing whatever of mining, and who judge of the capabilities of the captain by the amount of his cost-sheet. who, to their cost and to their pleasure, suits them (so he gets his salary), until the mine gets into difficulties, when they all declare it has been under careful management. Management! heaven save the mark. Management, forsooth! An order comes down to the mine, "You must limit your expenditure to 50l. per month: the committee will sanction no more." Management? What on earth is the poor captain to do after such an edict as this? His salary, of course, must remain intact, so must that of the purser, as well as of all the heads of the really expensive staff. The retrenchment falls on the hardworking portion of it—the miners.

In no case would we, if possible, transgress the *via media*, but adopt the miner's simple motto—"Fair tawing." In carrying out any extensive business operations mutual dependence, in a certain degree, must be placed in every branch of the employment. In no case is this more necessary than in that of the workman to whom we have alluded in the title of our paper. Having frequently heard adventurers grumble at this subaltern's pay exceeding that of the regular miners, and others called by that name, we have selected his case for illustration, that we may show, in depicting his duties, though paid at a higher rate, he is not overpaid, and that cheap labour is not always the most beneficial to the employer.

The most intelligent, sober, well-conducted men only are fit for or eligible to this important and responsible situation. It is absolutely necessary that he combines the practical experience of the thorough miner with the mechanical acquirements of the carpenter and the smith, and have a natural taste for and adaptation of mechanics. In short, he should be a working civil engineer. On him devolves the arduous duty of going to every part of the mine, to see the captain's directions are properly carried out, and to assist in their being done; to examine the security of all the precautionary measures, to watch any signs of decay or inefficiency in any part of the underground machinery or fittings; in short, the safety of the people employed in the mine devolves in the second degree upon this officer. Should any accident occur he must be there, if any danger arise he must face it. He is also expected to foresee such if there be any probability of its occurrence. We know an instance lately in which a fine fellow of this class had the moral courage to tender the resignation of his situation because he was not allowed to fix pitwork of sufficient capacity for the influx of water he saw was incvitable if the works were prosecuted. The result showed the man was right; the influx took place, and the mine was nearly destroyed. Practical miners consider the services of a thoroughly efficient pitman invaluable. The agents feel confidence, the men security; both these are of supreme consequence in such a dangerous occupation, where men's lives literally hang but on a thread.

From this class of men many, indeed most, of our intelligent practical agents are derived. They, in the plenitude of their success and elevation, boast with pride, "I was pitman at Wheal ——, many years before I was agent." That's the way to make men, and to understand mine work. Let 'em learn how to fix a column of pumps, put in a H-piece, change a bucket in five minutes, up to their middle in water, when there's no time to lose; and that is the way to make sharp fellows." All this, and more, he frequently has to perform. In fixing new work in old shafts his is a dirty, dismal, dangerous, situation. He has to direct and assist the fixing, and to stand by and order. When he has a lot of men, not miners, to manage, his is anything but an enviable position. Poor fellow! he has not only to exercise his own ingenuity, but his patience at their stupidity and clumsiness, the latter the more vexatious of the two. An old apothegm says, "You cannot make a silk purse out of a sow's ear;" nor can you make a good pitman out of a careless, stupid bumpkin. A manager, or a committeeman, may be made out of anything—even tailors and costermongers, by interest; but no interest will make the "Pitman." There is no danger of mines being brought to the verge of ruin by the agency of pitmen, though we have known them ruined by the former misplaced busybodies.

When men feel that from their own merits and endeavours they are advanced in their respective paths, they usually take heart, and endeavour still further to raise themselves. This noble sentiment and resolution has raised many a fine fellow to a proud position. Many a splendid mine captain speaks in glowing and grateful terms of the day he was made a pitman; and many an ardent youthful heart now beats in the hope of one day being advanced to this standing, which, if properly and duly attended to, is the stepping-stone to the "white jacket" (the mine captain's insignia), and from that to the "black coat" (the gentleman).

It will now be asked—Where is your purpose? Our purpose lies in a desire to see men rewarded according to their merits, and not according to their ostensible status or titles; to see pitmen paid according to their ability and responsibility, and not to have their wages grumbled at, as we frequently have heard done, by boards constituted, as we have said, to teach these parties to discriminate by the work, not the pay, whether or not they have miners in their employ; and not to dictate to captains what sums they shall limit their expenses, to, under the false guise of economy in mining. In no case is this folly more apparent than in employing a "cheap pitman."

Should this paper meet the eyes of those for whom it is intended, and they act on it, our end is served. The facts are known to many, and the actors evident to some old miners, and young committee-men, whom I conjure to recollect that cheap labour is not always profitable, nor a limited cost-sheet true economy.

Sir,—In your Journal of the 5th inst., under the head of "Cornish Mine Photographs—The Pitman," the writer says, "We knew an instance lately in which a fine fellow of this class had the moral courage to tender the resignation of his situation because he was not allowed to fix pitwork of sufficient capacity for the influx of water he saw was inevitable if the works were prosecuted; the result showed the man was right, the influx took place, and the mine was nearly destroyed." Now, Sir, I beg to ask, through the medium of your Journal, if Mr. G. Henwood means the pitman that recently resigned his situation at Holmbush Mine, near Callington? If so, I cannot allow this panygeric on the pitman to pass uncensured, on account of its invidious character, as it implies a want of due caution on the part of the agents. I am prepared to prove the assertions are utterly false, the pitman never having proposed to fix pitwork of sufficient or of any capacity or recommended any precautionary measures to be taken against the "influx of water" that "he saw was inevitable," other than was in operation by the direction of the agents; and, so far from the mine being "nearly destroyed," the water only rose about 5 fathoms above the bottom level, and this was in a great measure owing to the stoppage of the engine to repair a leak in the boiler, and other necessary repairs.

The reason assigned by the pitman to me for resigning his situation was, he said, because he thought he could get better wages on tribute, where he could take his son to work with him, than by keeping the situation that he then filled.

Mr. Henwood should be careful how he sacrifices the character of the agents of the mine for the purpose of exalting the "Pitman."

*Holmbush Mine, June 10.*                    N. SECCOMBE.

## The Mine Broker

IT WOULD ILL BECOME US TO OMIT THIS CHARACTER FROM
our catalogue of celebrities, forming, as it does, one of the most
prominent and distinguished features in the economy of Cornish
mining interests; it would be like performing the play of *Hamlet*,
omitting the principal character; we beg, therefore, to be allowed to
introduce him in this series of our papers. Like several others we
have previously portrayed, and many we shall, perhaps, at some
future opportunity attempt to picture, it must be considered in two
entirely different points of view; we, therefore, first proceed to
describe the regularly well-qualified gentleman, who ornaments his
profession by his ability and integrity, and who considers his business
an honourable and legitimate one, his own conscience dictating
those sentiments, and being, to him, a just reward—the *mens conscia
recti* that always has been, and always will remain, the patrimony
and true gratification of the really virtuous. Such there are amongst
mine brokers, many of whom are distinguished for their fair and
above-board mode of dealing; such as these require no "bush," or
boasting; they are quiet, silent actors, seldom seen or known to be
doing business, but who, nevertheless, transact extensively in expensive and really sound mines. These men are ever ready to give
information on a subject they really understand, and of which they
know the merits, provided the applicant requires it for *bona fide*
business matters, and not for gratifying an idle or impertinent
curiosity, as is too frequently the case. They carefully eschew display
or boast, and may be known by their quiet, unostentatious manners
and appearance, and would scarcely be supposed to daily transact
business to the amount of thousands; yet they do so without effort,
their quiet, business-like habits, being their best recommendation,
and the confidence of their clients their best support. Examples of
this description of broker may be met with in the metropolis as well
as in the country which has been chosen as the scene of our labours:
the distinguishing characteristics are the same, and may be easily
recognised, as they doubtlessly will, by those who have had
occasion to require their advice and assistance.

Solon, when asked "What was the most perfect state of society,"
is said to have replied, "When an injury done to the meanest
member becomes an insult to the whole community." Let us see
how this proverb applies under the present subject to mining affairs.

"Give a dog a bad name, and hang him," says the old adage;

breathe on a profession the breath of calumny, and you damn it, say we; yet it must in all fairness be acknowledged that the doings of some of the fraternity have given but too much reason for odium to be cast on its members generally. These unworthy scions are as easily known by their peculiar characteristics as their more worthy brethren—a degree of display, of loquacity, and a desire of having themselves distinguished in the market, in perfect contradistinction to our former portrait. We, as usual, shall not dwell on the unworthy portion; they are too well known to need particularisation— by their deeds shall ye know them. We can assure our readers the exceptions are few; the majority are as honourable a class of gentlemen as adorn any profession in the land; persons of education, of standing in society, and of real worth, whether considered as to their wealth or their virtues; whose word is their bond, and who receive the support and confidence of investors. Such gentlemen as these are above the tricks and cunning "dodges" resorted to by those who aspire to the appellation of mine broker, and herald forth to the world, in every possible way, that they, and they only, are capable of giving correct information and advice, but who, in too many instances, only delude the unwary, and ruin their clients, the greater part of them knowing nothing whatever of mining, and who look upon the sale of a lot of mine shares as they would upon that of a bale of wool or a yard of calico. To them mines and mining interests are as nothing, so that they get a shilling by transaction. This, to them, may be all very well, but it tends to bring others into disgrace, and mining into disrepute.

Instances of this nature, we are sorry to say, are too numerous for detail in the limits of our paper; we know they have been painfully experienced by many of our readers, and have frequently caused their utmost indignation; but we beg them to remember that they too have overreached themselves in endeavouring to procure bargains by purchasing cheap shares, with a view of becoming rich instanter, as if they were to be proprietors of Fortunatus's wishing-cap, instead of using patience and perseverance, the ground work and secret of mining success, of which facts they would have been informed had they sought the services of respectable mine brokers.

Another fertile source of the ill repute mining has undeservedly laboured under, is the vast humbug of foreign gold schemes, delusively called gold mining enterprises. These tremendous swindles were not entrusted to mine brokers, properly so called, as the media to place them before the public; a higher stand was taken, and they were called "stock," to the ruin of many, and utter waste of thousands of pounds. Nor was this all, it entailed a great evil on legitimate mining speculations—those capitalists who embarked, mistaking the shadow for the substance, included all in one fell category, and denounced mines, mining brokers, and all *sui generis*, a host of swindlers, schemers, and cheats.

Now we beg those who may labour or who may have suffered under any of these trials to reflect calmly, and ask themselves if they have not been at fault; if they were not, and are not, subject to the mania which periodically seizes mankind for the possession of suddenly acquired boundless wealth; and, if in most instances, the fault may not fairly be traced to this source? We ask them also to well consider the vast mineral wealth of Great Britain, and the consequences of its proper development in a national point of view; to reflect on the princely fortunes realised by well-matured mining projects, and the premiums paid by mining companies, superintended by men of ability and character. All these may be found, realised, and enjoyed by consulting and acting on the advice of a respectable "Mine Broker."

N.B.—With this character we purpose concluding this series of our Photographs, having portrayed a large number of Cornish characteristics as connected with Cornish mining. We purpose resuming the subject in a new series, which we hope will meet with as favourable a reception as has been accorded to that of which this forms the concluding chapter. Our only aim and effort has been to realise to view the real character of the mining community, and in this we trust we have succeeded.

## GREAT WHEAL BUSY

IT IS OUR PURPOSE IN THIS SERIES OF PICTURES TO APPLY our descriptions more to mines themselves than in that previously given, so as to afford a clearer view of the extraordinary value, and produce of these sources of national wealth, hoping by these means they may claim, as they deserve, a larger share of attention and support than they have for some years received from British capitalists. One reason for selecting this once valuable and extensive mine for our first illustration is that we must begin somewhere; and as this is now attracting considerable attention from a variety of circumstances, we think we could scarcely do better, or offer our readers a more acceptable subject with which to start.

The mine lies in what was once the heart of mining in Cornwall, and was the then fashionable locality (all districts become fashionable in turn if a rich mine is discovered), being literally surrounded by mines, some of which yielded enormous profits to the adventurers, and are well known as the Old and New Halenbeagle, Old Treskerby, Wheal Chance, Wheal Rose, East Downs, North Downs, Scorrier Mines, &c. The sett is situate about midway between the great granites of St. Agnes and Redruth, in clay-slate, traversed by elvan dykes (some of which are of great extent), as well as by metallic lodes, mostly having a run on their course for miles, in a direction varying a few degrees from east and west, having a northerly inclination or dip. Other caunter lodes are found also metalliferous, and frequently productive of copper ores; a glance at the strata would show the miner that it is highly probable tin and copper would be found in them. They vary in colour from a dark blue to a light buff, and frequently white killas or clay-slate. The elvans, too, preserve this peculiarity—they and the lodes vary in density and hardness, from that of extreme compact rock, or even iron, to so soft a texture as to be friable between the fingers.

On passing over the ground no one can fail to be struck with astonishment at the amazing heaps and burrows of stent (rubbish) which have been left by the former workers. They extend for miles; nor will their surprise be less when they learn that these mines, although they have returned hundreds of thousands of pounds profit, are absolutely scarcely tried; few of them have been sunk to a greater depth than 40 fms. Great Wheal Busy has been sunk no deeper than 100 fms.—an extent at which mines situate in a similar channel of ground would at present only be expected to yield extensive returns. This may be said to have been satisfactorily proved at the Consolidated Mines, United Mines, and others on the south, and not far distant from Great Wheal Busy. The practice seems to have been to work the mine as long, and where, ore could be found at a shallow depth, and then abandon the mine, without prosecuting discovery; this has been found to be an unsafe and unwise policy. Where cross-cuts have not been driven in productive strata a trial for side lodes cannot have been made; this has been the salvation of such mines as South Caradon, West Basset, &c. Now, in the instance before us, the mine is not at an unreasonable depth; the egg-shell has not been scraped out, as the present working, by prosecuting discovery, has proved indubitably. In the shaft sinking below the 100 a splendid lode, 6 ft. wide, has been found, consisting of a leader of copper on the northern part, 16 in. wide, by the side of which is a lode of tin of still greater value; the water exuding from this lode is quite warm, which is considered a most excellent symptom. In the 50 west also a most promising lode has just been found, yielding fluorspar (unusual in this lode), ferruginous quartz, black and yellow ore; this end is bidding fair to be an excellent one, and was found by a cross-cut. No doubt vigorous exploration would be equally successful; this, we understand, the mine is shortly to have, as indeed it deserves.

Too ardent and sanguine expectations seem to have been formed respecting this mine, and to have led to disappointment. A little reflection might have taught the adventurers that it was highly improbable the late workers would have abandoned a mine containing such surprising riches as were said to be there. The lodes have been found, but they had been stripped of their richest parts before the old men had left; still they are productive, as the sales of tin and copper ores show, but the quality has been found low; the ores found in the discoveries just made will greatly alter this; as, undoubtedly, when they shall have opened out room to work, valuable ores will be raised. These discoveries will not only be a great inducement to the present adventurers to proceed vigorously, but will enhance the value of their shares, as also that of the surrounding property.

The Hallenbeagle, once so very rich, and now generally supposed to be the best part of the mine, will, in all probability, be again tested, as well as Wheal Daniel (both these mines are in the Great Wheal Busy sett), new machinery being about to be erected for the purposes of the present mine, which will also be effective for these and the southern lodes; the influx of water being so great, the present engine would be overpowered were extensive explorations for discovery to be made. This is a step in the right direction on the part of the proprietary; their example and courage will, we trust, excite others to similar decisions, and not, as is much too frequently the case, neglect their mines just as the dawn of success is appearing, when they ascribe their loss to mining, whereas it is not to be attributed to that source, but to the want of perseverance.

This mine has suffered considerable vicissitudes in public favour: the present and future state of the property, we trust, will re-establish it to full confidence, not only for the sake of the present adventurers, but for the benefit of mining generally, tending, as it must, to dispel the cloud of adversity so long overshadowing Cornish mining interests. To this end all things seem now to conduce—a rising standard—the probability of more competition for ores—a beautiful harvest—cheap and abundant money, and a revival of speculation, as a legitimate consequence of these premises. Under all these combined advantages, we trust the Great Wheal Busy Mine will go on and prosper, and become a beacon to which all may refer when adverse appearances present themselves not to despond; for, in 99 cases out of 100, perseverance in mining, as in every other legitimate business, will amply be rewarded, even if the reward be a little delayed. We beg them to take heart, and say with the poet—

> " 'Tis not in mortal to command success,—
> We'll do more, Sempronius; we'll deserve it!"

In the *Journal* of Aug. 9, we stated our object in selecting this mine for illustration was solely its peculiar applicability to our purpose—to show the difficulties under which mining labours, and the unforeseen adverse circumstances to which it is liable, particularly in the resumption and reworking old and deep mines, from which, we trust, it will be seen that errors in the calculation of the costs are next to unavoidable, and should not in all cases be attributed to want of energy or ability in agents or promoters, as is too frequently the case, by over-sanguine and over-anxious adventurers.

The Great Wheal Busy Mine has been the subject of all these vicissitudes. We think when we have sufficiently and calmly viewed the matter, a proper allowance will be made by considerate adventurers, and the public generally, and tend to stem the torrent of invective so frequently indulged in if a speculation do not at once yield results, in time and value, as promised. Now, it must be admitted the glowing terms held out in the prospectus of this mine have not been fulfilled as yet—the extraordinarily rich ends have not been found; still a great deal has been done towards rendering the mine remunerative. To this end, five steam-engines, of the best construction, by the most eminent makers, have been erected, and are in a full state of efficiency; the largest engine, used for pumping the water from the bottom of the mine, is of the stupendous size of 85-in. cylinder, one of the largest in Cornwall, such engines have been found of sufficient capacity for draining many mines much deeper than Great Wheal Busy was reported and known to have been. It was originally intended to have erected three steam-engines, on different parts of the mine, for draining purposes, but the then agent deemed it prudent to try if he could not fork the mine with one only, when he would be able to form a more correct opinion as to the whereabouts for the additional machinery. Though this has been effected, the influx of water had been found so great, that the 85-in. is inadequate to its duty. Measures have now been taken to remedy this, by immediately erecting a 70-in. engine in such a situation as to command the other important portions of the mine. In addition to this, the recent discoveries in the deeper levels of the Great Wheal Busy Mine have given a stimulus to mining in the locality that will most probably induce others to work mines, and thus lessen the water charge most materially; indeed, the subject is on the *tapis*, and seems to be received with considerable favour. The other engines are adapted to drawing the ores and water from the mine, for which purpose two powerful machines are employed, each performing its work admirably. During the last week or two a skip-road, as it is called, has been completed, which will materially facilitate this operation; this, together with recent improvements, render this part of the machinery as nearly perfect as possible. The stamping-engine drives fifty heads of stamps, and is efficient for fifty more when required. The tin floors, burning-houses, &c., are laid out on a large scale, but are susceptible of improvement, which they most likely will gradually receive as the necessities of the mine may require, and as the force of circumstances may suggest. On these floors is a water-wheel, used for working a set of thirty of the self-acting frames, enabling a small boy to do the work of at least six or seven girls. Here, too, is a machine, invented by the late Mr. Herbert Mackworth, for dressing and washing coal, for which purpose it is admirably adapted, but, as far as tried, for tin-dressing appears to be a complete failure; at all events, it is but a complicated adaptation of Wilkins's separator, a much cheaper and simpler article, which I was surprised had not been applied at this mine. It is but fair to say the expensive article was put up at the cost of the maker, and not of the mine. Whilst on this subject, I must say I was also astonished not to see the round buddle and settling machines in use at the Great Wheal Vor and other mines; their cost is trivial, and may be easily applied to the present machinery at any time, as they doubtlessly will. The copper floors appear to be very well made, and a little time will so simplify the arrangements of the railways, &c., connected therewith, that they will also be perfect in their way. The crusher is a very good one, and does its work well; this is new, and is likely to be well employed. All this surface work has necessarily occupied considerable time, and a large outlay. The shears are made of cast-iron plates, rivetted together in the manner of the tubular bridge; they are the only specimen of the kind in the county, and answer the purpose as well as can be desired; they will surely be imitated. The rest of the surface arrangements are such as are usual, and call for no remark.

Underground, the mine appears to have been worked in the regular old style; every bit of ore taken away, except a few arches of inferior work. The lodes are large, the great lode (as one of them is called) runs nearly east and west, dips (as, indeed, is the characteristic of the lodes in the districts) north. A few fathoms from this is Winter's lode, which, from its having less inclination, forms a junction with the former; they appear to separate only at intervals, forming what are called "horses of ground." The lodes vary from 4 to 6 feet wide, but at the junctions it expands to 8 or 9 feet. We speak now of the western part of the mine, where the lode has been stoped; but still many pitches are set in the 30 fm. level, at an average of 12s. in 1l.; a considerable amount of ground may be let at this tribute, from which hundreds of tons of copper ore may be returned. The lodes are composed of capel, peach, and mundic, interspersed with spots and branches of ore, but is difficult to dress to a good percentage; the country in which they are embedded is a blue killas, of good quality for mineral. The deep, or county adit, here comes in at 40 fathoms from the surface; this adit has been driven upwards of 7 miles; 10 fathoms beneath this the lode comes into the shaft (Bennett's), which is sunk on its course, to the 50 below adit, and thousands of tons of copper and tin ores removed.

In the 40 tributers are working at 11s. 6d. and 10s. 6d. in 1l. for copper and tin. The lode seems to be composed of the same minerals throughout down to the 50, at the extreme west of which, the lode being very hard, and composed of capel, the level has been driving on the north part of it; a cross-cut south, however, revealed that there was a fine lode of fluor-spar, quartz, black, grey, and yellow ore: this is an important discovery, as it is all good saving work, and may be easily raised. The same description may serve for the whole mine down to the 70, where an influx of water is so great as to preclude working; this water comes from the west, is very warm, indeed hot, which is looked upon as an excellent symptom for proceeding in that direction. Men are now employed in the 80, driving to let this water down, when driving here will be recommenced; this is the water the new engine is intended to take up; at present it runs into the engine-shaft at 100 fms., consequently this stream (for it is as large as a mill stream) has to be raised 20 fms. needlessly; this will effect a very important saving. All these levels have had to be cleared, winzes sunk for ventilation, ladder ways fixed, ground secured, &c. This is accomplished; we merely introduce the workings of one shaft as an example of what has been done in the entire mine. The shafts and levels are too many to mention, their bare enumeration would be tedious, and not answer our end, as this is not intended as a report of the mine to the adventurers.

The great engine-shaft has had to be cut down at a very large expense to 100 fms. below the adit. As an evidence of the traffic there must have been underground in this mine, it may not be amiss to state that in many places, the shaft being on the underlie, there are no ladders, but steps are cut in the solid rock; these, from constant attrition, are soon so worn as to be completely rounded, the edges being destroyed; this, consequently, renders the ascent and descent of the mine difficult, if not dangerous, which will be remedied as soon as the agents have time to put all the place into proper and permanent working order. In the deeper levels of this mine are what are called carbonas—*i.e.*, excrescences on the sides of the lodes, bearing tin and copper ores; one of these, Moyle's bottoms, is of great extent, and shows no sign of being exhausted; as also a similar cavern, Chynoweth's bottoms. These excavations, as well as others of a similar character, though not containing very rich ores, have them in vast quantities, which will (when the arrangements are complete) be returned at a far less cost than when last worked, or even by the present appliances. The great feature of the mine, however, and one of vital importance to the whole district, is the fact of a change of ground in the bottom of the engine-shaft, where a splendid lode of tin and copper has been cut. We delayed the issuing of this paper for three weeks, to see if this discovery continued, lest we might have raised delusive hopes, and laid ourselves open to the charge of puffing a mine; we have now, however, the satisfaction to state that it not only "holds on," but improves in depth and is now worth 30*l.* per fm. for the whole length of the shaft. It is intended to sink 10 fms., and then drive on this fine lode, when the returns will not only be increased in quantity, but doubly valuable in quality. If this satisfactory evidence of improvement in depth be found in this place, what may not be hoped for in other parts of the mine? It should be borne in mind, many of our best mines have met such alternations about this depth, and have

ultimately yielded enormous riches, particularly if the large lodes
have been productive in the upper levels. In the 100 fm. level, or
sump, may be witnessed the curiosity of two mill streams, the one
from the west quite hot, and the one from the east quite cold,
pouring into the same pit, and from the same lode! In the eastern
part of the mine, owing to the recent improvement in putting in the
skipway, the shaft is not yet in thorough repair; this will soon be
rectified, and the mine placed in perfect working order; to effect
this a little time, a little more capital, and a little more patience must
be exercised, during which the recent discoveries will be developed.

In conclusion, we have only to say in depicting this mine we
adhere to facts, having no end but the benefit of mining *per se* to
serve. We implore others, as well as the Great Wheal Busy adventurers, to have a little patience, a little consideration, even if their
expectations be not immediately realised, when they will, in all
probability (as in this instance they assuredly will) have the satisfaction of witnessing their mines *de facto*, as well as *de nomino*—
Great Wheal *Busy* Mines.

## PORKELLIS MOOR MINES

THIS PLACE HAS OBTAINED A MELANCHOLY CELEBRITY,
owing to the distressing accident of last week, by which seven
persons lost their lives. Mishaps of this nature and extent are uncommon in Cornwall; though many men annually lose their lives
in the hazardous employ of mining, seldom more than one or two
work in immediate proximity of danger. This accident was wholly
unforeseen, and arose from the ground falling in from the surface
to a considerable depth, burying the levels and shafts, and, of course,
destroying the poor fellows employed beneath. At the time of its
occurrence upwards of 50 men were underground, and all who had
the opportunity instinctively rushed to the surface, the noise, which
they describe as being more terrific than the loudest thunder, having
given them notice of some fearful mishap. They at first imagined it
was an earthquake, their lights were at once extinguished, and the
horrible idea of being buried alive flashed across their minds. These
particulars we gained from one who was underground at the time,
and who barely escaped that dreadful fate. The extent of surface
gone down is about one-fourth to one-third of an acre, and is sunk

20 ft., as nearly as can be ascertained; the place now presents a deep pool of dirty water. To what extent the levels may be injured by the sludge and debris washed into them it is impossible to ascertain yet. So suddenly did the fall take place, that some girls at surface, who were working at the frames, narrowly escaped being buried. Some of the frames have gone down, and it is said several tons of tin, which had been prepared for the market, have been lost. Though every exertion will, of course, be made to recover the bodies of the unfortunate sufferers, some time must elapse ere that can be done. A large number of persons will be thrown out of employment. Fortunately three only of the men were married. The scene of distress and consternation is said to have beggared description, no one being able to form an idea of the extent to which the run might attain. The ground in the immediate locality is much shaken, and fears were entertained that the engines might be destroyed; fortunately this is not the case, but, in all probability, it will entail the stoppage of this part of the mine. The fall is within a few feet of the high road, from Wendron to Porkellis village, and it is a fortunate circumstance the road was not destroyed. The accident being the absorbing topic, we give these particulars previously to describing the situation and evident cause of the misfortune.

The Porkellis Moor has been worked as a stream work from time immemorial, and has been celebrated for the quantity of tin ore it has afforded—said to amount to millions in value; certain it is the vestiges of ancient extensive operations which abound, the old burrows and pits, extending for miles, attest it. Tradition gives a report of a small village near Helston having been the port where the Phoenicians traded with their small vessels for the much-prized mineral. Numerous granite millstones and mortars for pounding and grinding the ores have been met with by modern workers, some of which are of curious contrivance, and all bear the marks of long and severe attrition. Small parcels of tin ore, left by the "old men," are not unfrequent; and "jew's houses" (places where they melted their tin) have been observed. At a place called Trenear there is a curious building, evidently of very ancient construction, underground; it is built of well-wrought granite, and consists of three low rooms, the outer is about 12 ft. by 9 ft., the inner somewhat larger; in the former are niches constructed in the wall for some purpose not apparent, and there are no chimneys, from which it is inferred that it could not have been occupied as a dwelling. A large number of mills have been found in close proximity; these and the tradition would seem to give colour to the idea. Why the building should have been placed underground seems a mystery. It has been suggested this was done to protect the stores from the marauding bands who at that period infested the British shores; be that as it may, the building is still there for the examination of the curious, and is well worthy a visit by archaeologists.

The ancients appear to have wrought the tin streams on horizontal deposits, as this is the manner adopted on Tregoss Moor, Carnon, Nancothan, and other streams. There are, however, numerous lodes at Porkellis, which are productive of tin at the very surface; on reaching these, they sunk as far as them as the influx of water would permit, and for considerable distances on the course of the lodes. Similar works may be seen at the Kellyer's stream, near Halloon. These are exceptions, and not the rule. These "tinners' hutches," as they are termed, are generally large pools of water,—the drainage of the surface and rain falls,—and are very dangerous places. The Porkellis Moor is full of them, and they are sources of great inconvenience to the regular miner. The ground, too, is of a loose character, being growan or decomposed granite, some part of which, known as the great clay, is so highly dangerous as to deter workmen until the mine is thoroughly drained by having numerous levels. Lying in a pan of ground, or nearly a dead level, and surrounded by hills, it is not surprising that the accumulations of water are excessive, and require powerful engines for drainage. The cause of the disaster is, no doubt, the tributers having exhausted the veins and backs, until the ground left was too weak to bear the superincumbent weight; or that the timber put in for safety was insufficient in strength. This is a very delicate question, the growan having the quality of swelling when exposed to the atmosphere—this is called by the miners "plumming." To such a degree in some mines is this the case, that what in one week would be deemed perfectly safe, in the next might be highly dangerous. This we particularly noticed at the Wendron Consols Mine, adjoining Porkellis, where, however, ample precautions had been taken. The moors may be said to be one continual series of lodes, running nearly east and west. Besides the two mines mentioned, are Wheal Ruby, Caledra, Wheal Foster, and others not at work.

The present company have expended many thousands of pounds in developing the mine and erecting efficient machinery, and were in sanguine hopes of reaping the reward of their spirited outlay, when this untoward event has prostrated their efforts for a time. We sincerely hope, however, that their loss will be found much less than anticipated, and that they, like the Atlantic Telegraph Company, will not allow one mishap to damp their endeavours, but that they will have the magnanimity to rise superior to their misfortune, and by perseverance yet reach the goal of their exertions, to their own honour and profit. All concurrent testimony is in favour of the Porkellis making a good mine, as well as its neighbours, who have paid dividends, and are now in an excellent position. Had not this fearful calamity befallen them, it is highly probable the Porkellis would have been a dividend mine ere long. We hope yet to see it in that position, by practising a little energy, patience, and capital.

## CHACEWATER VILLAGE

THIS VILLAGE APPEARS TO BE A COLONY OF MINERS WHO have worked in mines in various parts of the world. Scarcely a family is to be found one member at least of whom has not been out either to Mexico, California, Brazil, New Zealand, Australia, Africa, Spain, or some mining district of less account; in many instances their wives have accompanied them. The Portuguese and Spanish is well and very generally spoken by them when conversing on the subject of their foreign experience. It is not only amusing but highly instructive to listen to the details of their trials by field and flood. From these it is to be gathered that the first who went out to Cuba to work the copper mines suffered far greater loss by death than later emigrants to that country—partly owing to the inexperience of those early visitors, and partly to the carelessness of the miners themselves. Many have returned a second time, and some remained in the country for fifteen years and upwards. Nearly all secured a little competency, to enable them to get into some way of business, a public-house or beer-shop being the principal and favourite speculation. Some few have realised sufficient to maintain themselves in a state of independence. Nearly all the officers on the foreign mines are Cornishmen, and, from the representations of the miners, appear to be admirably conducted and carefully wrought. They also speak in high terms of the native Spanish descended workmen, particularly the Mexican portion, who they say are quite equal to the Cornish. Loango, in Africa, seems to be the spot most dreaded. I met a most intelligent captain, who had been out there in quest of malachite, in which he was pre-eminently successful; shattered health obliged him to return, and no temptation on earth would induce him to go out again. Two youths who have just returned from the same station, mere wrecks of mankind, give the same doleful description; still, if men are required, no difficulty in procuring them is experienced, the wages offered being too great a temptation to be resisted. Those who have returned represent the mines as being extraordinarily rich in malachite of the hardest and finest qualities, and, could the physical difficulties of the country be overcome, inexhaustible wealth could be procured.

In close proximity to Chacewater some mines of great celebrity have been worked, including Scorrier, North Downs, Wheal Rose, Treskerby, Hallenbeagle, Great Wheal Busy, Wheal Seymour, Wheal Daniel, Creegbrawse, the United Mines, St. Day United, Consols, &c., returning many hundreds of thousands of pounds profit, and employing a very great population. The entire village, and much of the neighbourhood, is the property of the Earl of Falmouth. When the mines were in their palmy days Chacewater was a place of considerable importance as a mining village; a capital market-house was built for the convenience of the people, but has since been almost deserted. As the mines became abandoned Chacewater fell into decay and poverty. By the spirited endeavours of a few individuals the Great Wheal Busy has been set to work, where upwards of 700 people are employed—a great advantage and blessing to the locality. Although this great undertaking has not yet made profitable returns to the adventurers, its promising appearance and satisfactory progress has stimulated adventure, and many mines have been, or are about to be, put to work by powerful companies. Within a radius of four miles of the place many millions of pounds worth of copper and tin ores have been raised and sold, so that to speak of the productiveness of the strata would appear a work of supererogation. Suffice it to say that it is of the slate formation in the neighbourhood of the granite, and traversed by elvans, hornblendic and porphyritic dykes, as well as by metalliferous veins of all descriptions found in Cornwall.

The village, as a consequence of the resumption of the mines, is reassuming its former *status*. Several good substantial shops have been occupied by a superior class of tradesmen; and two or three good inns are to be found, mine hosts of which do all in their power for the comfort of their guests. Other shops are being erected, but the stringent clauses of Lord Falmouth's leases (all the property is built on leases of three lives) militates against the building a superior class of houses. The places of worship are well and regularly attended, and the ministers much respected. A cricket club, and freemasons' lodge of no mean pretensions, have been established; indeed, the place presents all the elements and appearances of a thriving mining village. Long may it continue, and be emulated.

The town, too, has all the peculiarities belonging to country places. Everybody knows everybody's business better than their own, and act accordingly; this will cure itself as the population increases and business becomes more active, when they will have "other fish to fry" than talking scandal. Here, too, may be found the sharp, shrewd, witty, and persevering tradesman, with a joke and a kind greeting to all he comes in communication with. Here, too, is the half-witted butt; and the intelligent, kind, and respected doctor, known to all and knowing all, "breed, seed, and gener-ation." Here, too, is the connection in business, engendered by family marriages and relationship, that almost prevents the success of any stranger endeavouring to establish a trade; this, however, will be remedied by the extension of commerce, necessarily following the exertions at present being made to prosecute the Chacewater mines.

The markets are amply and cheaply supplied, and the people well employed, contented, hospitable, and well conducted. We should like to see many villages equally prosperous and promising as the village of Chacewater, the centre of a great mining district. Though her sons may be occasionally deported to work foreign mines, they frequently return to their homes laden with wealth, improved by mixture in superior society, and the experience travel always engenders, the beneficial effects of which cannot fail being imparted to all with whom they are connected, having, as experience shows, an elevating, self-respect creating tendency, as well as a finer appreciation of the benefits by which they are surrounded, and is an inducement to thankfulness to Him by whom all these blessings have been accorded.

## COOK'S KITCHEN MINE, ILLOGAN

THIS ANCIENT MINE IS PROBABLY ONE OF THE MOST remarkable instances of the continuity of Cornish mines if per-severingly wrought, having been in constant work for 100 years. In its earlier days it was celebrated for the production of tin ores. It is also said to be the first place at which copper ore was appreciated; at that time the adjoining mine was called Bullen's Garden (now Dolcoath). Documents still on the mine date as far back as 1766; they are highly valuable, and exceedingly interesting. About 14 years since, by an unaccountable act of vandalism, nearly half a ton of these memorials were burnt by the then agents of the mine.

The old books preserved throw great light on the circumstances and mode of management at that time practised. From a notice in one of their pages we find the mine was worked by a "rag and chain pump," instead of the ponderous and costly machinery now used. The record says—"15 August, 1766: Borad of Capt. Haten Donken a ragedbehead with all hooks, and 34 foot of chain."—"2 Sept., 1766: Borad of Penhallack (now Tincroft) oners 14 lbs. rope." —"Do. of do., 15 lbs. of rope." Though the orthography is so very imperfect, yet the accounts appear to be clearly and regularly kept, the settings of the various pitches being specified in a manner well worthy imitation by modern captains. It also appears that the custom of "spoling"—that is, fining the miners for dereliction of duty—was practised. For the month of October, 1766, 41 men (of whom six were punished in this manner) and 38 girls were engaged in dressing the ores, whose wages amounted to 14l. os. 6d., or at the rate of about 4d. per day. The list of names contains those that are still familiar to mining people; many of the men now working on the mine are descendants from the old worthies. One very singular entry appears as "Christopher Bennats a horse skin." This material is now not used in mining; it was possibly applied for thongs, or traces, for the whim horses, as I perceive Bennats was a whim driver. It is shown that at the end of every month the materials not consumed were returned by the men, as there are entries as low as 2 lbs. of candles, and 1 lb. of powder. Subsist was, as now, given to tributers, of whom 14 were employed. We insert two specimens of the mode of setting:—

"Cook's Kitchen marks for November, 1766: Westran bottom, William Stephens taker— 1 mark is a abonyar hole in the south wall, near the cros cose, 2 foot from bottom; 2, is a boyar hole in the south wall, 2 foot from bottom; 3, is a boyar hole in south wall, 2 foot from bottom; 4, is a boyar hole in south wall, 3 foot from bottom, and 2 foot from the end: Meyard, 29 Novemb., 1766."

"1 Novemb., 1766: A survey hild at Cook's Kitchen Mine, for setting the north wins Brea lode, to sink, drive, and stope as directed, to six men until the 29 Novemb., 1766; and to save the ore carefully, or be spolid 2s. 6d. each time that such neglect shall happen, and to be conformable to all the customs of the mine: sett to Wm. Tellam for 15s. per fms."

At this period the mine was making extensive returns, considering the number of hands employed, as we find the following entries:—

" Cook's Kitchen ore sampled 7 August, 1766:—
   Computed tons, 20; waid 20 tons 12 cwts. 0 qrs.; price, 18*l.*  5s. 6d. per ton.
   Computed tons, 10; waid 10 tons  0 cwts. 0 qrs.; price,  8*l.* 11s. 0d. per ton.
" Sampled ye 2 Oct.:—
   Computed tons, 23; waid 23 tons 13 cwts. 2 qrs.; price, 17*l.*  9s. 6d. per ton.
   Computed tons, 19; waid 20 tons  7 cwts. 0 qrs.; price,  5*l.* 10s. 0d. per ton.
   These lots were bought by Benallack and Smith."

From this period to 1790 the mine seems to have made rapid progress, as on Feb. 4 of that year the sales amounted to about 2000*l.* for copper; the tin bills and prices have been destroyed. The agents' salaries were 2*l.* 10s. per month, and the miners' earnings were about 1s. per day. Amongst the singular charges for sundries is one to Jane Prideaux, for setting forth the corpse of Mr. Arthur, 5s.; Wm. Arthur, consideration for his son being killed in this mine, 1*l.* 1s.; and Mary Bennetts, for an eight-day clock, 5*l.* 8s. The doctor's money amounted to 1*l.* 15s. 6d. per month; and the purser's to 3*l.* 3s. In this year the costs amounted to about 1000*l.* a month. Until the month of October no mention of the impost of samplers' fees occur—in this there is an entry of 10s. 6d. for the samplers; and in November there was a meeting on the mine, at which the victuals are charged 3*l.* 18s. These charges from this time seem to have been kept up, as in 1797 Mary Cocking was paid 9s. 2d. for victuals supplied on setting and pay day, and 1*l.* 7s. 1d. for ditto on the account day. The first entry of ticket expenses appears to have been on Nov. 4, 1797. when 4*l.* 13s. 3d. was charged, and 10s. 6d. for samplers' fees, but nothing appears for weighing off the different parcels. In December, 1797, Boulton and Watt's savings for this month are charged at 18*l.* This would imply the use of an engine; the particulars of this machine and its cost are unfortunately not to be traced, or they would be of great value. I feel little doubt these particulars will be highly interesting to many of your readers, who may not have an opportunity of perusing the original.

What a mighty contrast does the mine now offer to its earlier stages, where we find about 30 or 40 men and girls at work on the entire mine; at present more than 200 are engaged, assisted by four powerful steam-engines, four water-wheels, and all the appliances of modern inventive ingenuity, which enables the same expenditure of manual labour to complete 20 times as much as could have been accomplished in olden time. Every improvement is adopted that science suggests, from the time of Boulton and Watt to the present. Here are Teague's dressing machines, which are worked by a simple contrivance,—an adaptation of the old flap-jack,—instead of water-wheels, these are very inexpensive and efficient. At the depth of 22 fms. from the surface is a water-wheel 50 ft. high, formerly used to pump out a part of the mine.

This property has, like most others, seen its vicissitudes. It gave 144,000*l.* profits in 12 years, and paid 1-12th dues to the Basset family; since that period a loss of 30,000*l.* was entailed. The mine has, however, been again rendered remunerative, and at present is realising a profit, with every probability of continuing to do so. The greatest depth attained is 250 fms., and the quantity of tin obtained is 14 tons per month.

With this notice I conclude the Photograph of the Cook's Kitchen Mine, to the agents of which I owe much for the courtesy and readiness with which they afforded me access to the documents, and furnished me with such information as I required to complete this paper.

GREAT TYWARNHAILE MINE

IN OUR LAST JOURNAL WE GAVE EVIDENCE OF WHAT PER-severance in mining could accomplish, and showed a mine to have been continuously wrought for upwards of 100 years. In the instance before us we, unfortunately, present a striking contrast. This great property, after raising and selling a vast quantity of copper ore—in two years amounting to 16,935 tons,—was abandoned, the then low price, nearly the lowest ever quoted, being the cause; the owners not having the courage to further prosecute the adventure. Their outlay was undoubtedly great, and the prospects of Cornish mining most discouraging; however, scarcely had the materials been disposed of when the market for copper ores advanced to very nearly double the value. Many miners were likely to have shared a similar fate from the same cause. Had not their adventurers "waited a little longer," they had not seen the, for them, "good time coming," which many have had ample reason to rejoice at, for their endurance was rewarded by discoveries of immense value, when the price was, as it still is, of a fair remunerative standard. At the same time of its suspension, many far-seeing personages declared the mine to be a sacrifice indeed, and many were the arguments, *pro* and *con*, on the subject, the *Mining Journal* and the Cornish papers being the vehicles through which remonstrances of the most powerful nature were conveyed; amongst these the well-known pen of "Argus" (of Truro) was one of the most distinguished. Still, so great was the pressure of the times, nothing availed, and that which might have conduced to the welfare, wealth, and comfort of so many human beings was abandoned to desolation and ruin. Where the valleys once echoed with the clatter of machinery, the busy hum of industry, the deep-toned voices of the miners practising their hymns or carols, or the jocund laugh of the "bal maidens,"—some hundreds, in the aggregate, being employed on the mine,—nothing is now to be seen or heard but bleating goats, two or three mountain sheep, with now and then a stray passenger, who ejaculates, as he passes over the wild heath, and views the stupendous wreck, "What a pity! what a pity!" when the echo, or if he has a companion he echoes, "What a pity!"

For these sympathies there is very good reason; they are very extensive, and pervade all ranks, from the highest to the lowest in the county. But to none are they so painful as to the practical, professional miner, to him witnessing this exceedingly valuable property lying dormant is a source of the most sincere regret; he knows and feels for the misfortune of all concerned, and is also aware of the capabilities of the great mine, which, by vigorous exertion of competent persons, and an adequate outlay of capital, might be made a property available to an extent scarcely credible to a novice. All testimony tends to this conclusion. Since the stoppage, science has so improved the machinery for the manipulation of the ores, as well as for exhausting the water in the mine, and the value of the produce has so far improved, as to make it a matter of all but certainty that the Great Tywarnhaile must succeed if re-worked.

With a redundance of capital in the country awaiting profitable employment, we confess our astonishment that this property has been overlooked or neglected by speculators. One or two ineffectual attempts were made during the late stagnation in commerce to recommence the mine, but the panic prevented any effectual measures being adopted. Now, however, the case is materially different; and we hope and trust this Photograph may be the means of directing attention to this mine. It may be relied on as correct, and written with the view solely to benefit the adventurer as well as the locality, feeling assured, as we do, the prosecution of it would be equally beneficial to the one as the other. The mine is the property of the Duke of Cornwall, has buildings of great cost and extent already erected, and levels and shafts driven, of course causing a large saving to any future company that may be wise enough to embark a sufficient capital to secure a large and profitable return. We hope then to be enabled to see the Great Tywarnhaile with much greater pleasure than we have done in our late visit, and to again witness the scenes and hear the sounds we have previously described, knowing them to be the true signs of industry, wealth, peace, and contentment; a consummation devoutly to be wished, and one that is to be easily and quickly attained.

We say, in conclusion, that we purpose portraying several of the Cornish mines in the manner we have attempted this and some others, feeling the true representation of their merits and circumstances will do more good, and be more conducive to the success of our home mining interests, than any other means we can select. When depicting the riches of some of our great mines, it must not be set down that we are overlauding their capabilities; or when we describe them as they are, be supposed to be detracting from their merits; as either case is wholly foreign to our purpose. We hope to publish some interesting, instructive, and curious details; in doing so we shall cultivate variety of locality and subject as opportunity and necessity may suggest.

## CARN BREA HILL

DURING OUR RAMBLES AMONGST THE MINES IN THE immediate neighbourhood of the "sacred hill" of the Druids, we, on a beautiful autumnal Sunday afternoon, wandered to the top of the mountain, and after examining what are frequently considered the relics of these ancient priests, in the shape of altars, rock basins, &c., we seated ourselves at the base of the Dunstanville Monument, and fell into a speculative reverie of the past to the stern and palpable realities of the present; from a fanciful picture of the dread procession of the human sacrifice through sacred groves, and all the horrible concomitants of the worst form of pagan superstition, to the quiet, holy aspect of the Christian Sabbath, the stillness being only broken by the sound of mine stamps (these ministering to the wants and comforts of man, instead of the sounds devoted to a foul and horrid idolatry) wafted occasionally on the breeze. We thought of the vast difference civilisation had wrought, and of the wisdom of Him who directs mundane affairs, as exemplified from the situation on which we sat; for assuredly had man the power, his cupidity would have long since exhausted the vast resources of this, perhaps, the richest spot on the face of the globe. We then set about considering the vast sources of employment afforded to a redundant population by the very difficulty of procuring the necessities mankind in its progress cannot do without, and to this end proceeded to count the different mines within the scope of our view. We fain would have done so, but far off, and further still, rose the dim outline of some engine-house or chimney stack, till our patience became wearied. We could distinctly recognise and name upwards of 100, on all of which we had been, and knew their productions and peculiarities. We then cast our eyes on the broad ocean, which from our altitude appeared much nearer than it really is, and we numbered no less than 132 vessels. We pondered again on the wisdom and the power that

"Spread the flowing seas abroad,
And bade the mountains rise."

There we sat, on the summit of this hill, in the centre of some of the richest mines ever worked in this or any other country. There, at our feet, lay the Carn Brea Mines (so called from the mountain), with its numerous family of East, North, &c. Carn Breas. There away west, Tincroft, Dolcoath, Cook's Kitchen, and Stray Park, all of which have been wrought for generations without sign of exhaustion. On a parallel with these, to the north, lie the Seatons, the Roskears, the Croftys, the Pools, the Tolguses—we are compelled to use the plural in their case as well as in our own; for true it is, that if one mine in a locality obtain distinction by the wealth it may produce, the name is adopted without any real claim to the nomenclature; and the cardinal points of the compass, as well as the cardinal features of truth, are frequently outraged to accomplish the desired name. This is wrong, and should not be practised. It is too frequently a delusion and a snare, and as such should be guarded against. But we digress. Confiding ourselves to the subject-matter of our paper, we turn round to the east, and our vision is greeted by the sight of the Treleighs, the Pedn-an-drea, the Cupids, the Gramblers, the ancient Sparnon, the famous Bullers and Bassets, the Old and New Penstruthal, the glorious old Tresavean, the pride and boast of Gwennap, notwithstanding her famous Great Consols or United Mines—all these may be seen, by a mere twist on the heel. Another move displays the Frances, the Rosewarnes, and at the distance of a few miles radius may be discerned the whole of the western mines. Oh! a glorious sight is it from the top of the Carn Brea Hill!

And from all this what is to be deduced? This—Is British mining profitable to this country or not? The question has many able advocates on both sides, and is a theme on which a vast amount of ability may be exercised. We have heard the question argued most cleverly, but were the debaters to be seated at the base of the Dunstanville Monument, we think that silent monitor—conscience—would decide that the evidence of the senses is the best and most overpowering that can be adduced; that the millions of pounds worth of metals extracted from the mere propinquity of the mountain are so much national wealth absolutely found, and are positive gain. Let they who dispute consider, as we did, how many cargoes of timber have been used in these mines—how many thousands of ship loads of coals, tallow, and copper have they been the means of causing the transport. Look at yonder fleet of vessels, many of which doubtless, at present are freighted with produce to and from these subterranean sources of industry. But hark! there's the whistle—the railway whistle! We doubt if the sound of that would have ever been heard in this solitude, or the iron net of Great Britain ever reached its *ultima Thule*, had it not been for the mines within our ken.

The mighty steam-engine, the handmaid of science, the friend of labour, the modern Briarcus, if not born, has been nurtured, cherished, and perfected (as far as it is perfect) within view of this spot. Yes; let us see. There is Cosgarne, where the immortal Watt once studied; there is the place the genius Trevithick was born; there is the residence of Woolf; there is the witty and clever Vivian, the companion and friend of Trevithick; there is the very road on which the first steam locomotive ever ran; there is the residence of Hornblower; there of Murdoch, to whom posterity owes so much. These are the Nature's noblemen whom mining has fostered. Look, see now, on yonder group of trees to the north-west, with the mansion just peeping—that is Tehidy, that Carclew, that Trewince, that Scorrier, the residence of the millionaires, who owe their all to the persevering industry exercised in the caverns beneath the landscape at our feet. The neat white mansions and park-like lawns, scattered about in profusion, are the abodes of the smaller adventurers or managers of these mines, bespeaking an amount of comfort and ease not to be excelled in any county in Great Britain, or in the world. Oh! a glorious sight is it from the top of Carn Brea Hill.

Now that the Cornwall Railway is so nearly approaching its completion, and when so many visitors will be within reach of the old Druid's Mountain, we implore them to visit—ponder—reverie, as we did. To ask in the presence of all we have advanced, is British mining profitable? Had it not been so, whence all this scene? Its permanency from time immemorial to the present is indubitable, its future incalculable; of its present the view is sufficient evidence.

Whilst dwelling on this subject, we should not forget the Obelisk, at the base of which we sit. Here it stands, a noble, grand, and enduring monument of the men who work these mines. No prouder example graces the memory of man than this, for it is a token of sincere gratitude, a memorial of true feeling, alike honourable to those who built it as to him in whose honour it was erected. It is the devoted offering to the manes of him who was really the miners' friend; it is not like Pope's—

" —— tall bully lifts its head and lies."

No inscription need be placed on this edifice, it tells its own tale; we, therefore, shall not further dwell on it.

But time flies apace, evening shades come o'er us, and we must descend; in doing so we must be careful, as the path is difficult. At length we reach the high road, swarming with well-clad persons of both sexes, bespeaking comfort and happiness, wending their way to church or chapel, into one of which we enter, and find one of their own class ministering to his fellows that comfort he has the language and talent best to convey to them; for had they the eloquence of a Tillotson or of a Macaulay it would not be half so acceptable, or so well understood, by those unlettered sons of toil. We retire to our hotel, highly gratified by our long walk and longer reverie, and retire to rest with a full conviction that "it is a glorious sight from the top of Carn Brea Hill."


## REDRUTH MARKET DAY

AN ANCIENT WRITER SAYS—"TELL ME WHO YOU ARE WITH,

and I will tell you what you are." This is only an axiom, and not a truism as is generally supposed. The witty Sidney Smith observed, that "If a man be born in a stable, he need not therefore be a horse." Diogenes, when seeking in the market-place at Athens, was not necessarily a "Cheap John," though it is proved he carried a lanthorn. Nor are we at Redruth doing more than a cosmopolitan visit, to observe the ways of men now at Redruth, and compare them with the former days, "when we went gipsying, a long time ago." Another author says, "The best criterion of a people's welfare is to study the comforts they possess, and the means they have of obtaining them." This is true political economy, for if the mass have the means of common comforts, the aristocracy have easier minds; if, on the contrary, the people have not the facility of obtaining the necessaries of life, and that too, by ordinary exertion, something must be wrong, and vows not loud but deep will be engendered— "A leary belly makes a saucy tongue."

We are led to these observations by spending a day at Redruth, and contrasting its present state to what it was in our youth, some forty years since. It chanced to be on a Friday, the market day at this now really fine town. After considering its present flourishing condition, we reflected whence all this? To this end we called on several shopkeepers, making trifling purchases, for the sake of enquiry, when we invariably found them regretting the good old times "what was," and waiting for the good time coming. True is Pope's line—"Man never is but always to be blessed." We now but apostrophise; we must to our purpose. Well, to begin at the beginning. We look out of our hotel window, and see two robust tradesmen in the street, one a knight of the cleaver and the other a boniface. We unwittingly hear the common observation of "Fine day, Sam." "Iss, Sir; I believe we shall have a good market. Tin is rose, and the standard was up yesterday. They do say that Carn Brea is cut rich, and Buller is better in the bottom. We shall have good times again in Redruth yet." At uttering this his countenance was lit up with good humour; and the innkeeper's jocund assent testified his gratification, as he sped back to his hostelry of the Red Lion.

We laid down our razor without effecting our purpose, to reflect on the scene open before us; and, wending our way to the spot we had determined to be the scene of our day's observations, noticed first the supply and suppliers of the articles on sale. These we found so multifarious, and from so many parts, that, like the bellman of the town, who too was overwhelmed with business, declined reading the particulars of the handbill he was engaged to describe, as being "too numerous to mention." We found, however, that the prodigious quantity of comestibles, in one shape or the other, were derived from all points of the compass. That whereas thirty years since about ten butchers frequented the market, more than thrice the number now retail; that everything had advanced alike; that no such thing as barely flour was used. We then enquired, how is all this? when we were responded—"The mines, Sir, the mines." Here, then, was the grand secret—"The mines, Sir, the mines." Here was the golden key, the solution of all the comfort, and the source of all the wealth, prosperity, and happiness of which Redruth Market was such palpable and convincing evidence. Hark! what's that cry,—"Sheers! Sheers! Sheers!" We walk up to a curious conveyance, a pair of shafts, with a cross piece for the wheels to act on, and a couple of baskets (panniers) filled with the delicious pilchard and less luscious but equally grateful hake. This was the advent of a fish supply, equal in quality and variety, if not in number, to Billingsgate, the language of which celebrated community has been communicated by degrees, and beautifully less, even to this remote scene. This portion of the market is, shame to the authorities, carried on in the steep and narrow thoroughfare, to the annoyance of everybody, whereas a little vigorous effort, as at Penzance, would abolish the nuisance, without encroaching on "vested rights" in the slightest degree.

Here come the butchers, with their well-filled carts, robust, healthy faces, and clean blue frocks and sleeves, depositing, jointing, and putting the best side of all pieces "towards London." It is curious to witness the *modus operandi* of a Redruth butcher, to see how carefully he apportions a bit of bone with every joint, and to watch how nicely he weighs out every portion of fat with the lean. That part of it which with the London butcher best suits the chandler is here most in request. Yes; the tallow is the most valuable part. On enquiring why, we found it was used by the mining housewives for making pastry, in the shape of haggans, pasties, &c. "Haggans!" we hear exclaimed, "what are haggans?" Haggans, we are told, are haggans. But, ho, stranger! if ye list, oh, list! to hear of a dish which even Soyer, that prince of gastronomics, never thought of—"pilchards and leeks in a pie!" Oh, ye who never knew the joys, try it! Remember Redruth Market, there you can have both in perfection, as well as all the butcher's meat in due season, and in no town in the kingdom is it in greater abundance or of better quality.

Go we now into the streets, where, as I have before said, the nuisance still exists: over that we have no power of description—it is absolutely infamous. But here is our old friend Cheap John, from Sheffield, with his everlasting cargo of the last lot, and his eternal stock of stale lies and obscene jokes. Here, too, is our old, very old friend, the worm-doctor, with the veritable leather boot-string in a phial of cold water, representing in full the character of its charlatan master; being the true representation of the "black tape-worm," which, as he loudly exclaims, "eats a man's vitals out." Here comes the last dying speech and confession man. Here the verses composed by one of the late sufferers at Porkellis Mine, now published for the sake of his afflicted family, price only one halfpenny. Oh, a rare market is Redruth! But, make way, here comes the king—no, the mayor in state. "Lo! he comes, from clouds descending." But here he is, and we must dilate upon his good qualities, as in duty bound. Well, his good qualities are patience, open ears and long, willingness to do any duty, under any kind of treatment; his only bad quality is obstinacy, and where is the potentate that is not ass enough to have that.

We walk through the market of well supplied vegetables and all other delicacies the most addicted *gourmand* could desire, and then we wonder how is all this profusion to be consumed? We visit the hotel. There we find Capt. Friggers shaking hands with Capt. Wilkins, still in earnest conversation with Capt. Pope; Capt. Pope having hold of Capt. Rogers's button; Capt. Rogers holding up his stick to Capt. Richards, who was hallooing to Capt. Cock not to go to bal till he had seen him. Here, thought we, is the great secret of all this bustle—Mining.

We enter the long-room, as it is called, filled with smoke, mine captains, brokers, greenhorns, &c. Our advent (being known) is the signal for general, not universal, salutation. "Well, old fellow!" resounds on every side. We sit down and enjoy ourselves for awhile —we are prone to fraternise; old Virgil, when describing a storm, said, "*Nil nisi pontus et aer; nubibus hic tubimus, fluitibus ille minax.*" This was a similar vortex, nothing but mining, or the price of wheat; all as it should be for a market; farmers and miners each driving his own, though very different, wheelbarrow. We, as in duty bound, having quaffed our malt, visit another hostelry; the self-same subjects though in different hands, with a little dissertation, by way of amusement, on the prizes of the St. Leger, bull-dogs, and tin stealing, with a little local scandal by way of spice; as the latter became plenty we became scarce, and found ourselves at mine host's, where horse topics were uppermost for the time, which soon gave way to mining. We retired with the full conviction that mining only was the proper subject for Redruth; that the all-absorbing, that the parent, that the supporter, and that will be the end, of Redruth.

That mining may long be its glory, pride, and source, and that many more may emulate it, is the sincere wish of the miners' friend.

## Carn Marth Hill, Gwennap

THIS ELEVATED SPOT IS ONE OF THE GREAT GRANITE bosses of the supposed substrata of Cornwall on which the clay-slates, and other metalliferous rocks, are superimposed. Its outline and characteristic are not of that abrupt and picturesque description which render the Carn Brea Hill so attractive to strangers; it is therefore, less visited by them for that reason. We do not purpose dwelling so much on this part of the subject in our paper, as on the mining instruction to be reaped by observation from its eminence; though much might be gleaned from Carn Brea of the Illogan Mines, the Gwennap series on the east are of a different character; notwithstanding they produce similar metals, the minerals and strata in which they are discovered differ materially.

This hill, or miniature mountain, is quite as high as the Carn Brea, and commands an equally extensive view. From it may be seen the Bristol and English Channels, giving a tolerably clear idea of the narrow track of Cornwall's peninsula at this part. It is a particularly barren spot. Though surrounded on all sides by mines of considerable magnitude, and of unheard-of, incalculable value, its granitic base, unlike the Illogan rock of that name, has never been wrought to advantage, and has, therefore, got so much into disrepute as to deter further trial. From it, however, spring the series of elvan dykes which are supposed to have such powerful influence on the stanniferous and cupriferous lodes of the Gwennap and Kea Mines. The elvans are of two varieties, hard and soft, and very like granite in appearance, but on examination will be found to be importantly different; they are, however, favourite rocks with the miner, and in this locality abound; the hill may be said to be the great centre of mines, as our description will, we hope, display.

We were accompanied to the summit by a friend who had never ascended it before, though he had been visiting the Cornish mines for some time; it was, therefore, necessary to point out the leading features of the landscape to him, as well as to describe and name the different mines within view; in doing so, our description will, we trust, serve as a correct guide to our readers.

Within a radius of two miles from the spot on which you now stand, more than 20,000,000*l*.—aye, than 30,000,000*l*.—worth of minerals have been raised from the bowels of the earth, and have been wrought for generations. That narrow winding valley on the south-east was the scene of, perhaps, some of the earliest endeavours at mining ever practised in this country; its origin has been lost in oblivion; but certain it is the Carnon Stream, for that is its name, was wrought at a very remote period, possibly before the present race inhabited this island. Down this valley at your feet must once have rushed a mighty torrent, hurling a prodigious quantity of tin before its impetuous flood, thus providing an easy mode of pro-curing it, when man, from his imperfect knowledge of metallurgy and gun-powder, would have been unable to pierce the tremen-dously hard rocks he even now finds it difficult to explore. Wonder-ful Providence! Striking illustration of order and contrivance in the works of the Deity! In working this stream ample evidence was discovered of the primitive races who sought tin here. The imple-ments of work were wooden shovels, in a few instances shod with iron, showing how valuable a metal iron must then have been. Many picks formed of the horns of deer were discovered, as well as other contrivances of a rude construction. The ingenuity of the present age has not only enabled the tinner to rework the old men's refuse at a great profit, but to pursue his avocation beneath yon tidal river, where you observe a number of shipping. Yes, beneath the navy there, conveying coal to and carrying off the produce of the mines, works the tinner, at a depth of sixty feet, in silence and safety. That is the port of Devoran, the *entrepot* of the mines of this district, by which, indeed, it was entirely created. That railway, whose locomotive you may perceive runs to Devoran, is of vast advantage to the neighbourhood—watch it; see how it threads through the different mines with its immense burden!

This group of engine-houses are the surface buildings of the United Mines: let us count—one, two, three, aye, seven engines in a small space. Those beyond, in the same line, are the Clifford Mines. Still further east are the Great Baddern, Wheal Jane, and

East Falmouth. On the north of, parallel to and in immediate proximity, are the Great Consolidated Mines, with their ten enormous engines. On the east are Nangiles and Wheal Sperries, Wheal Whiddon and West Wheal Jane. Now, if you observe the run of the burrows, you will perceive the lodes of these mines run a little to the south of east and north of west: when we descend I will show you their dip, or inclination from the perpendicular. But look further north; that group is the St. Day United Mines, beyond is the Creegbrawse and Penkivel Mines. Still further north, and parallel, are the Wheal Jewel, Wheal Damsel, and West Damsel; that very extensive mine on the still further north is the Great Wheal Busy. The village, whose church you may perceive, is Chacewater; that nearer is St. Day. The scattered engine-houses you see still further north are in the St. Agnes district, and comprise North and South Ellen, the Great Tywarnhaile, Wheal Towan, &c. Besides these, on the hill you see Polbreen and Polberro. The hill is St. Agnes' Beacon, a place we purpose visiting, and passing its peculiarities in review.

There is Redruth, which, with its vast multitude of mines, we have already described. On the north-west is our old site, Carn Brea; that on the west Carn Menelez Hill, another boss, around whose base tin mines abound; but as that is out of the subject matter of this paper we leave it. There is the Penstruthal; there, further south, is Wheal Comfort; that on the hill, just on the south-west, is Tresevean; that mine at our feet is Old Ting Tang; beyond it, Wheal Moyle and Wheal Squire. These vast heaps of debris are the indications of the extent of these subterranean excavations, at the bare idea of the cost of exploring which the imagination staggers, and makes the uninitiated shudder.

Having taken a lengthened survey of the most interesting subjects within our view, we descended the hill to visit the principal mines, and make enquiries by which we might illustrate the subject of our discussion on our pedestrian journey—Does Cornish mining, after all, conduce to the national weal, as well as to the local exchequer? To this end we visit the United Mines, taking them as an average of the mines in the Gwennap district, which, we were informed, were a fair criterion. Before doing which we proceeded to the spot at which we promised to show the underlie, or dip, of the lode; this we found at the base of the Nangiles Hill, where the miners have stoped the lode to the surface, and the vein is denuded. This great lode is from 14 to 20 feet wide, dips north 20 deg., is composed of capel, jasper, quartz, prian, chlorite, oxide of iron, mundie, copper, and tin ores, and is the *beau ideal* of what a lode should be. Being wrought to the open day, the *modus operandi* of mining may be here readily and correctly understood by those who have not physical power or nerve to descend a mine. We know not of an example affording so practical and favourable an illustration; it is well worthy a visit, and is easily found. Our friend was delighted with the opportunity of studying so fine a lode as this by daylight.

The UNITED MINES have been wrought from time immemorial— at first for tin, subsequently for that metal and copper ore. In these mines is the celebrated "Hot Lode," the water issuing from which is 114 deg.; so hot indeed is the atmosphere that the men work perfectly naked: cold water is brought from the surface to pour over them, but even then they can stand in the end but a short time at once. This part of the mine is very rich: from the shafts communicating with this the steam rises in so dense a body as to look like smoke from a chimney. The agents of the mine obligingly favoured us with the following statistics, which we believe are as nearly as possible correct, though the number of persons employed and the supplies vary, as a matter of course, very considerably at different periods; we could only obtain particulars from the time at which the present proprietors took possession. The first sale was in May, 1851.

*Amount of Ores sold in Money Value.*

| | | | | | |
|---|---|---|---|---|---|
| 1851 | £25,895 | 9 | 1 | | |
| 1852 | 55,334 | 6 | 1 | | |
| 1853 | 63,084 | 7 | 10 | | |
| 1854 | 54,593 | 7 | 5 | | |
| 1855 | 58,418 | 8 | 7 | | |
| 1856 | 51,981 | 9 | 6 | | |
| 1857 | 48,647 | 18 | 1 | | |
| 1858, to Oct. | 37,635 | 9 | 0 | =£395,627 15 7 | |
| Deduct lord's dues, rates, and taxes | | | | 20,000 0 0 | |

| | | |
|---|---|---|
| Leaving | £375,627 | 15 7 |
| Dividends* | £25,000 0 0 | |
| Present market value | 32,000 0 0 | =£ 57,000 0 0 |
| Outlay, 400 shares at 40*l.* | 16,000 0 0 | |

Profit in seven years..... £ 41,000 0 0

*Consumption of Materials, Labour, &c., Annually.*

| | | |
|---|---|---|
| Coals | £12,000 | 0 0 |
| Candles | 900 | 0 0 |

| | | | |
|---|---|---|---|
| Timber ........................ | 700 | 0 | 0 |
| Gunpowder and fuse ............ | 900 | 0 | 0 |
| Iron and brass casting chains, &c. .. | 800 | 0 | 0 |
| Hemp ropes, &c. ................ | 1,000 | 0 | 0 |
| Machinery ..................... | 500 | 0 | 0 |
| Labour and salaries ............. | 27,000 | 0 | 0 |
| Horse hire, carriage by rail, & quay dues ................. | 1,000 | 0 | 0 |
| Sundries ...................... | 500 | 0 | 0 = £ 45,300 0 0 |

\* To this item should be fairly added the sum of 10,000*l.*, said to be unnecessarily paid to the neighbouring mines as water charge.

On such data as these we could but come to the conclusion that Cornish mining does not only contribute to the general wealth of Great Britain, but that it is a most important section of it; that without it the Cornish gentry and landowners would be impoverished to an extent incalculable; that the benefits of the dues of the Cornish mines are extended to landowners of the county resident in all parts, and amount in the aggregate to an enormous sum, in procuring which they do not incur the slightest risk or expense, and is, there-fore, clear gain to the country at large; that commerce and shipping thrive to a vast extent by mining enterprise; that the supplies consumed in the mines are derived from all parts, and must necess-arily yield an equally divisional profit to each person engaged in them; that British mines tend to civilisation, comfort, and happiness, and that without their presence England would lose her power, and prestige amongst nations; and Cornwall, instead of her present active, teeming population, and highly prosperous state, would have been a howling wilderness and a desert, her sons barbarians, and her gentry paupers—in short, that the mines of England are her main stay and safeguard, and that in this glorious category the mines of Cornwall form a principal and leading feature, and as such should be cherished and encouraged. The mines and miners deserve it, and God grant they may receive it.

ROSEWALL HILL

"FISH, TIN, AND COPPER," IS A PHRASE THAT HAS BEEN A standing toast in Cornwall for centuries, and is as duly used at feasts of all classes of persons and professions as any of the loyal or patriotic toasts of the day. When it was first adopted is unknown, but the arrangement of the words would imply it was a period anterior to the produce of copper superseding that of tin. The pilchard, mackarel, and other fisheries, along the extensive seaboard of Cornwall, have long ministered to the necessities of mankind, and, with the tin streams, were probably the chief reason of Cornwall being comparatively densely inhabited at the early period we find recorded in ancient history. The true position of the relative interests have, however, greatly changed; indeed, quite reversed. Probably from no eminence in the county could we more appropriately study or illustrate the ancient and lasting staple produce of this richly-endowed county; in doing so, we shall follow the modern system, and place the interests in their true position, as "Copper, Tin, and Fish."

The Rosewall Hill is a granite boss of considerable elevation, from which a fine view of St. Ives Bay, and all the neighbouring mines, may be obtained. It is surrounded on three sides by trap rocks and compact argillaceous slates, which have yielded, are yielding, and probably will for many generations continue to yield, vast quantities of copper and tin ores. It is the junction of rocks to which this peculiar fecundity of mineral is usually ascribed. Within a radius of two miles are some of our most productive mines, the lodes of which have been proved to produce alternately tin and copper ores, and not unfrequently so intimately and equally blended as to render it a matter of difficulty to pronounce for which metal the stone is the more valuable. In this particular they coincide with other celebrated mines, whose geological features are synonymous—as the Botallack and Levant Mines in St. Just; Dolcoath, Cook's Kitchen, and Carn Brea, near Redruth. We have in some of our previous papers said copper ore was first wrought for in Cornwall at one of the latter mines; we find, on enquiry, that we were right.

The extensive mine at the foot of the Rosewall Hill, although worked as a tin mine, has frequently made considerable sales of copper ore. It is called the St. Ives Consols; is a very extensive and celebrated mine, having returned vast quantities of tin, and paid the fortunate proprietors hundreds of thousands of pounds profit. In this mine was discovered the largest carbona, or floor of tin, ever found. On describing the Providence Mines, in one of our previous papers, we published an engraving, and detailed description of these phenomena, for they may justly be called so, being peculiar to this locality. To those who may not have seen our paper, we beg to describe them as being huge excrescences from the sides or walls of the lodes, having no particular shape or outline; the lodes in which they are usually found contain hollow places, called "vughs" by the miners; a good carbona is usually a great fortune. This mine, in its most palmy days, was under the management of the late Capt. Tredinnick, father of the present Mr. R. Tredinnick, of Austinfriars, and in this very mine did this gentleman practically lay the foundation of the mining knowledge that has served him in such good stead.

We digress, we are aware, but incidents and facts like these should not be overlooked or forgotten. The mine between the St. Ives Consols and the position we occupy is the old Ramsom and Rosewall United Mines. These have been wrought to a great extent by various companies, and at good profits; the rich tin ground trending away west from the St. Ives Consols promises a rich harvest to the present adventurers; who, from the energy displayed, and capital invested, certainly deserve success of the first degree. On the north-east, to the left of St. Ives town, is the Carrack Dews Mine. This is in a hard killas rock, produces little tin, but holds out high promise for copper ores: here may be procured the curious and beautiful varieties termed *horse-flesh* and *bell-metal* ores. The mine due east of St. Ives Consols is the Wheal Trenwith, once exceedingly rich for tin and copper produce, the latter preponderating—fortunes were realised, and the mine abandoned. A few years since it was set to work by a party, who literally carried out the old nursery distich—

> The King of France, with twenty thousand men,
> Went up a hill—what then? Came down again.

This they certainly did—they set the mine to work with a powerful steam-engine and effective plant, pumped out the water, and abandoned the mine without sinking it a foot, or giving it a trial; the mine at the former working gave a profit of 40,000*l*.: the mine is in killas. To the south-east of this, and near the cliff, is the Wheal Margery, which produces both tin and copper, but is usually reckoned a copper mine. This property is regarded as an excellent investment, the returns gradually increasing in quantity and quality as the mine gets deeper; it has not hitherto paid dividends but will probably do so during the current year. This mine is near the junction of the granite and killas.

Turning further south we perceive, near the monument of the late Mr. Knell, on the top of a hill, the Trelyon Consols Tin Mine. This mine has returned considerable amounts of rich quality tin, and has paid good dividends. The mine is not deep, and is entirely in the granite formation.

Further south is the Providence Mine, which has been worked for many years; the lode may be seen in the cliff, and copper ore broken therefrom. The mine has yielded large quantities of copper ores, but is now only worked for tin, of which it contains what may be at present termed an inexhaustible supply. The mine has returned about 100,000*l.* profit, and pays regular and handsome dividends. This mine is full of carbonas, none of which occur above the 60 fm. level: the copper ore was procured in the clay-slate, and the tin is in granite.

Immediately joining is the East Providence Mine, lately set to work under the most promising auspices, and by a powerful company of capitalists; recent reports hold out high promises of success, which we trust will be verified. The Providence Company have worked their mine nearly to the boundary of the two setts; and the East Providence party entertain high hopes of the riches of the parent mine being continued into their property. On the south and south-west are the rich tin mines, Wheal Margaret, Wheal Kitty, the celebrated old Wheal Reeth, from all of which immense dividends have been derived. On the west of the latter mine are the Reeth Consolidated Mines, from which more than 200,000*l.* worth of tin has been sold during the last twenty years.

Looking away west as far as the eye can reach, traces of old works and streams are visible; no record exists of their being wrought; but as all this part of the country is "bounded" (a right of possession to certain portions of the produce, peculiar to Cornwall), it is pretty certain these lands have anciently been extensively searched. Between the Rosewall Hill and Wheal Reeth is the Balnoon Mine, once in high favour, but now abandoned. The newly-erected engine-house and works on the north-western declivity of the hill, are of the Brea Mine, now attracting considerable attention, as discoveries of importance are reported. This mine, is like the Providence Mine, St. Ives Consols, &c., at the junction of the granite and killas, are said to contain both tin and copper lodes. The mine at present is in high favour; and certainly, if position amongst mines, or peculiar strata, be any recommendation, this property contains the elements of prosperity that can hardly fail.

We need not indicate which way to look for the beautiful bay of St. Ives, with its silvery sands and bluff black cliffs; from our heights they are laid out as a map at our feet, forming part of as exquisite a panorama as can be seen in any part of the county. Far off, on the east, on a fine day, may be seen the entrance to Padstow Harbour, nearer, those to the port of Newquay and Portreath; almost at your feet the Bar of Hayle; these form the northern ports, whence Cornwall's produce is shipped to Wales, and to which coals are imported therefrom.

This celebrated bay is the principal seat of the great Cornish pilchard fishery; the quantities taken in some years surpass belief, in others scarcely a single fish is secured: the fishery is quite as great, if not a greater, speculation than mining. That spirit of adventure, so leading a characteristic of the Cornish people, however, supports this interest with amazing zeal. Though the "schools" (shoals) of fish pass through the bay but rarely in a season, yet so eager are the inhabitants to have a part of the prize, that notwithstanding only about six or eight boats can fish at one time, and each of these within its prescribed limits, as if the fish go from one "stem," as the distance so prescribed is termed, they cannot be followed by others, each must wait its turn; thus only a few hours can be allotted to every boat. An Act of Parliament was obtained for the due prosecution of the fisheries, by which the limits became defined and legalised, yet not less than 150 "seines," the name of the net in which the fish are enclosed, are engaged to secure the finny prey. When taken they are prepared in a peculiar and coarse manner, packed in cases called hogsheads, containing several thousand fish each, and are sent to the Levant, and many ports of the Mediterranean, where they are highly prized. The oil expressed from these fish is valuable to the currier, and is much in demand. The quantity taken at St. Ives during 1858 amounted to 12,000 hogsheads, of the value of about 3*l.* per hogshead, irrespective of the price of the oil, which is usually calculated to pay the cost of curing the fish. The aggregate of the other Cornish fishing stations may be said to be about equal to the preceding. Thus, it must be admitted, Cornwall derives no inconsiderable revenue from this one species of fish. The mackarel fishery is of equal value; and the general coasting, or white fish, as that sort is called, may be considered as representing a similar amount. These fisheries have been practised from the earliest ages, and undoubtedly were once considered her principal staple commodity.

It is only by ascending such heights as we are supposed to be on that we can form a just appreciation of Cornwall's extent, or of the amazing sources of wealth developed by her industry. From this spot may be seen the places where thousands daily ply their toil beneath the soil or on the treacherous ocean, each in his sphere adding to the nation's wealth, and to domestic comfort and happiness. It is well worthy of a visit by the naturalist, the geologist, the political economist, or the general enquirer.

We think, by these premises, we have shown why the old toast was adopted in the manner expressed, and that we have given ample reason why it should be reversed, and "Copper, Tin, and Fish," be substituted.

## The Carglaze[11], or Open Tin Mine

AMONGST THE MANY INTERESTING SCENES WITH WHICH Cornwall abounds few will at once prove so entertaining to the archaeologist, the geologist, or the mere casual observer, as this celebrated mine. The history of its commencement is lost in obscurity, but the vast size of the mine itself affords abundant proof of extreme antiquity, whilst the numerous works at surface and immediately adjoining, which have been wrought exclusively for extracting tin from the backs of the lodes, evidence the origin of this prodigious undertaking. The mine is situated on the apex of a high hill, about $1\frac{1}{2}$ miles from the beautifully situated and thriving market town of St. Austell, to which place access will shortly become very easy, as one of the main stations of the Cornwall Railway will be there. The circumference of the shaft, or entrance to this mine, if measured in all its eccentric sinuosities, would probably exceed a mile. At first view, from the edge of the abyss, it presents the appearance of an immense bowl-shaped excavation, save that it partakes of an oval instead of a round form, its length from north to south being about a quarter and its average breadth an eighth of a mile, and is 20 fms. deep. The stratum in which the tin ore is found is a kind of decomposed granite, or growan as it is technically termed, traversed by great numbers of small veins, or lodes, of tin and schorl; these may be easily detected in the sides or walls of the mine, which being of a colour nearly approaching white, display them to great advantage, as these minerals are of a blackish colour. The lodes vary in their bearings, some being nearly vertical, others having various inclinations or dips; the majority of them have an east and west direction, or strike, but a little attention will enable the student or observer to trace caunter veins as well. Most likely there is no place in Cornwall where the practical examination of tin veins by the novitiate can be made with equal facility; indeed, we know of no spot at which a "field lecture" on the subject could be so advantageously given. The great extent denuded, the privilege of examination by daylight, the depth at which the lodes may be traced in their descent, the junctions and the splitting of mineral veins here to be seen *in situ*, are opportunities which can scarcely be over valued; but, strange to say, they do not appear to be thoroughly appreciated by students in geology.

To the admirers of the picturesque a rich treat is afforded. The former workers left many parts remain that were not rich enough to pay for removal; the effects of the weather, as well as the pick of the miner, have converted these blocks into the most fantastic forms and groups, presenting in some instances sharp peaks and sugar-loaf miniature mountains; others are like huge masses of snow; the whole are so variegated in colour by the oxidization of the minerals contained therein by atmospheric action, contrasted with the dazzling whiteness of the precipitous cliffs at present working, form an unique *coup d'œil*, which is further heightened by the ruins of several abandoned stamping mills which were long since used for crushing the ore raised from the mine. These are now overgrown with lichens and mosses of every variety of colour, and abound in the most luxuriant and graceful fern foliage. That necessary appendage to all landscape beauty, water, is to be seen in a bounding stream of considerable volume, running for about half the length of the excavation. This to a stranger presents a singular appearance, being as white, and of nearly the same consistency, as milk, from the charge of china-clay it receives at the present works. This stream was anciently used to wash out the tin ore and work the stamping mills before alluded to. The value of the china-clay at that period was either unknown or unheeded. Be that as it may, it is now sedulously preserved. An adit, or subterranean tunnel, has been cut, at an enormous expense, to convey the liquid through the side of the hill into receptacles of several acres in extent, for the purpose of deposition, solidifying, and drying for the market. The process of preparing china-clay is so similar in all the works, that a description of the method practised at Carglaze will serve to illustrate them all.

At the upper end of the works the growan, or wall of the hollow, is upwards of 60 ft. deep: this growan, or decayed granite, it is scarcely necessary to observe, consists of felspar, quartz, and mica. The felspathic substance is purely white, and is the portion so highly valued. This growan is dug by workmen, and is subjected to the constant action of the stream before mentioned, by which the felspar and mica become separated; the quartz, tin schorl, and other mineral substances are at once retained by their weight, whilst the felspar and the mica are conveyed by the stream through the tunnel to the first series of depositories, termed mica pits, in which, from the superior specific gravity of that substance to the fine clay, it is precipitated, and the felspar conveyed to numerous other fields or pits, of which the last contains the finest qualities. These are filled to such an extent as to afford a depth of about 1 ft. of clay when dry. The water, when perfectly clear, is pumped off, and the rest desiccated by the atmosphere. The clay is cut into balls, or lumps, of about 1 ft. square, is rendered perfectly dry, by being placed in open sheds; after which it is denuded of all extraneous and fatuitous impurities, and is then the china-clay, or kaolin, of commerce, and is shipped to nearly all all parts of the world; for, exclusive of its extensive consumption in the manufacture of earthenware and porcelain, it enters largely into the composition of writing and printing papers, bleaching and stiffening calicoes, paper staining, &c. No use is at present found for the refuse mica, though doubtless science will eventually realise the desideratum.

At present the Carglaze is worked for the clay alone, the small quantity of tin discovered not being anything worth mentioning. The uncertainty of mining may at any moment reveal a discovery of importance. The limits of our paper preclude any further description of the interior, though we have dilated on not one-half of its attractions; we, therefore, ascend by the crooked and steep pathway we descended.

If the visitor be fortunate enough to be on this eminence on a fine day, a most extensive prospect may be enjoyed, and one scarcely to be equalled in extent or variety. On the north are the great china-clay works, the most extensive in the world, from which upwards of 20,000 tons are annually exported. Here, too, is the peculiar variety of semi-composed granite, termed china-stone, of which about half that quantity is raised. In the extreme distance, on the north-east, may be distinguished Cornwall's two highest mountain-like hills (Rough Tor and Brown Willy); on the east the rugged Luxulian and the Great Caradon hills, with their rich quarries and mines; beyond is the Kit Hill, with its monument-like mine chimney on the top.

Further, on the south-east, is the entrance to Fowey Harbour; near the obelisk visible on the promontory are the Fowey Consols, so justly celebrated for their prodigious returns; nearer to the spectator is the district of Par Consols and St. Blazey; the Port of Par, wholly constructed by the late Mr. Treffrey, may be plainly distinguished, as well as the little harbour of Charlestown, whence the produce of the ground on which we are now standing is shipped; the middle distance is made up of mines that have realised enormous fortunes for their lucky proprietors, amongst which properties may be enumerated Pembroke and Crinnis, Great Crinnis, Boscundle, the old Apple Tree, and the St. Austell Consolidated Mines; the aggregate of whose produce in tin and copper would stagger belief, amounting to millions! Looking due south, the visitor will be delighted with the enchanting view of Charlestown Bay, apparently at his feet, studded as it at all times is (except in gales of wind, when we presume no excursionist will visit the Carglaze) with numerous coasters, fishing boats, and shipping from the ports of Charlestown, Fowey, Par, &c. The extreme point on the left is at the entrance to Fowey Bay; that on the right is the Dodman, or extreme point leading to Falmouth; within this bay are fisheries of considerable extent. At Polperro and Mevagissey the pilchards are frequently secured in large quantities, but never to the extent mentioned in our paper relating to this subject in connection with St. Ives. On the south-west is the town of St. Austell, whose prosperity depends entirely on the clay works and mines in the locality. The extensive mines on the west are the Polgooth, Great Hewas, St. Austell Consols, and Dowgas Mines. These, with those before enumerated, comprise the St. Austell district. Far off in the distance may be seen many others, the notices of which have formed the subjects of previous papers. No person can ascend these heights without being struck with the number, prominent position, and beauty of the church towers; from this place a large number may be observed. To judge aright of the extent, variety, and worth of this celebrated county's productions such visits are absolutely necessary; the mere casual observer does not see, nor can he appreciate, the extent and value of the St. Austell clay and mining district accurately, unless such a pilgrimage be taken, and to no better point can it be directed than to the subject of our Photograph.

## Allotting Shares in the Olden Style Revived

SOLOMON SAYS "THERE IS NOTHING NEW UNDER THE SUN," but had a stranger witnessed a scene we recently beheld he would have certainly deemed the sacred sage in error. We would we had the descriptive powers of a Dickens to portray the scene; but even that being the case, it could be but faintly delineated: the reality surpasses the most vivid pencil, or most ludicrous caricature. In our boyhood we had heard of such doings, but never witnessed, and never wish again to witness, the like. It proves the old Cornish spirit is not dead; it is still vigorous as ever. Our readers may well almost doubt the truth of our picture; we pledge our honour it is literally true, and could we but present the grotesque *dramatis personae* the effect of their introduction would add doubly to its value.

The mine was "got up" solely as a Cornish Company, by a spirited mining captain, who had been abroad, and had returned with a considerable sum, the result of his persevering endeavour and toil.

The magic of wealth, or supposed wealth, as a matter of course so universal, was not exceptional in this instance. The capitalist soon applied for and obtained the lease of a valuable mining sett in the immediate locality. Having the money to set to work, men were employed to clear up the old workings; lodes were discovered; a shaft was sunk on a supposed lode—a really splendid gossan, with rich copper ore, soon found. The proprietor soon installed himself as Capt. R., chief manager of the mine *pro tem*. A well qualified secretary was invited to join the speculation, who readily embarked in it, introducing a large number of powerful and influential parties into the undertaking. The mine being surrounded by some of the richest mines ever worked, whose returns are almost deemed fabulous (amounting to many millions), soon attracted considerable attention, particularly as the discoveries were deemed so eminently satisfactory by all practical miners who had visited the spot. This was its state at our first visit to the mine, and we vouch for its accuracy. From this time, daily visits to the rude little wayside inn of vehicles of every description evidenced the interest the mine was exacting. Applications for shares literally poured in—were begged for, and many were they who would gladly have taken 100, but who could procure no more than 10.

But the course of human affairs never did, and probably never will, run smoothly: there became two parties, owing to a quarrel and want of "compatibility of temper," we think the Divorce Courts term it. So it was at this mine; the two interests—the captain's and the purser's—exerted themselves to the utmost, and with the most virulent and determined hostility all the while.

At length it was resolved to hold a meeting, the rival interests each insisting on his right of dictation and management. On the appointed day a dinner was to be held on the mine, for which each claimed a right to and did cater, and well it was so, as the sequel will show. Duplicates of everything were provided, and each party invited his friends and patrons as guests. On this important day, when the future management of the mine was to be decided, hours before the time appointed crowds of pedestrians, equestrians, and vehicles of all sorts might be seen hurrying to the scene. Groups were to be noticed traversing the various parts of the sett, and descanting on the probabilities of success, but all regretting there was a division in the camp. It seemed universally accorded that the mine itself was a fair speculation.

There being no mine buildings large enough to hold all of the assemblage, adjournment to a little inn hard by became absolutely necessary, the largest room of which was soon crowded to suffocation.

The secretary announced that no shares would be recognised but such as were there and then produced, and the deposits paid; on this, the whole number not being forthcoming, it was resolved that the shares of the company should consist of the number now applied for, and that no more be issued. The meeting then proceeded to examine accounts, to appoint officers, &c.; but, ye Gods, what a scene! The jargon at Babel would have been comparatively intelligible; both parties, and all at once, urging their respective claims to respect, at the same time vilifying the other; vociferating, gesticulating, and making confusion worse confounded; creating such delay, that it occupied five hours to accomplish what one hour's calm reasoning would have effected—a reconciliation and reappointment of the agents in their respective situations, and a determination to erect the engine forthwith.

The company were then requested to leave the room, that the cloth might be laid for dinner. But not so. A seat at the board was too valuable a possession to be vacated. The double invitation had congregated a crowd of at least 150 hungry men, most of them miners, farmers, and tradesmen from the neighbouring towns, who had been sharp-set from 8 a.m. "No, no!" was the reply; "bring here the articles, we'll soon lay them," which, in truth, they did; not one-half could be accommodated with seats, much less knives, forks, or plates. Huge pieces of beef, legs of mutton, hams, pies of all descriptions, and of gigantic proportions, soon vanished as if by magic. The rich merchant, the surgeon, the adventurer, the invited and uninvited guest, had each to carve for himself, and struggle as he best could for a seat or a scramble. It is no joke to spend ten hours on a bleak down, at the distance of five miles from a town, on occasions like this. It was suggested that as one party dined, they should retire for the next. "No, no! Where's the grog?" A seat was again far too valuable a possession to be vacated. The poor purser's patience was sorely taxed; for had he for one moment vacated his seat, the place had known him no more. Every room in the place was filled to repletion.

Decanters of spirit were placed on the tables to be but emptied in a trice, refilled again and again with a similar result. This scene of revelry continued until many could literally carry no more, and the mob began to disperse. The quantity of spirituous liquors provided for the occasion though ample and liberal, was, as the miners say, soon in "fork;" additional gallons were obtained from the inn until their stock was exhausted, and two gallons had to be fetched from an inn at a distance of a mile or two. A more incongruous mass never before congregated, or one which was more difficult to control. However, all ended well; the purser got his calls well paid up, and, but for the double *contretemps*, all would have been well. The jolly old Cornish custom of a good dinner and good friends constitutes a good company and creates good and ready payments of calls, but on this occasion out-heroded Herod; instead of being what it should have been it was more like a bear garden than a feast, and a drunken revel than a meeting for business.

Happily such scenes, as we have before said, are now extremely rare; we had heard of such things in the olden time, but scarcely believed them possible; we saw it, and, therefore, record it as an extraordinary and rare occurrence. It is scarcely to be credited that at a mine meeting at this day more than 12 gallons of spirit, independent of ale, porter, and quantities of lemonade, &c., should have been consumed, yet this was done, and declared by many to be the good old Cornish mining times coming again.

Gentle reader, let us now pause, and consider that all this worse than useless extravagance arose from that very source which has been the utter ruin and destruction of so many, and that the expenses must be borne by the mining companies themselves. Quarrelling and division were the cause of this scene of folly, recrimination, and delay. Instead of promoting the interests of the mine it militated against it most materially. It teaches, or should teach, promoters of mines that indiscriminate invitation is not true hospitality, or lavish expenditure the true means of securing respect.

We had reserved this paper until the novelty and fame of the well-known meeting had settled down into partially a matter of local history and scandal; but we assure our readers that such scenes were once quite common, and the rule; we rejoice to say they are now the exception, and point to this as a beacon and warning against quarreling and party spirit, which leads to equally great, if not more serious, excesses.

## THE "OLD MEN" AND THEIR WORKS

THE APOTHEGM, "OLD MEN KNOW YOUNG MEN TO BE fools," is partially true, as is its paradox. "An old man and a fool, too." In mining, the latter adage is too frequently applicable; sometimes an addendum not quite so modest or complimentary might be used—we refer to the saying more R. than F.

How often, gentle reader, have you met with a cantankerous old miner, who knows everything but his business—this he usually professes to know better than anybody else; who, because he has age to protect him, dares to utter opinions in such an offensive style as in a junior would insure castigation. His ideas are propounded in such bombastic, egotistic terms, that the word "fool" is strictly correct, and the epithet rogue equally true, for if he be the one, depend on it he is the other; in fact, the antithesis of modesty and truth.

The mining *genus homo* of "old men" may be properly divided into two distinct series—the ancient old men, and the modern old men; the former distinguished by the magnitude of their works, the latter by the hyperbole of their sayings. The ancient old men have in cases innumerable left sterling proofs of their industry, perseverance, and ability, so as to have and deserve the passing tribute—"The old men were good miners, after all," to which we profoundly say, Amen.

The object of this paper is to direct attention to some of the more remarkable of their works that have fallen under my ken, as being in connection with modern mining pursuits, and by these means induce a visit to them by young miners, when if attentively observed will surely yield instruction, amusement, and profit. One of the most remarkable instances is to be found in Porkellis Moor, parish of Wendron, Cornwall, where the remains of the abnormal process of procuring tin may be found covering the entire valley; here also may be seen to advantage the same process repeated a second, or even a third time, the old men only taking off such rich produce as would pay the cost of carriage, smelting, &c., at that time undoubtedly very expensive items; remains of "Jews' houses," or places where the ancient old men smelted their tin on the spot by charcoal, are sometimes met with, when refined tin in singularly shaped blocks or ingots is discovered. These are very rare, and highly valued by antiquarians. I have seen one example where the action of the ground, which was highly charged with oxide of iron, had transformed the metallic tin into its original state of ore, thus plainly indicating that the growth and decay of metals is continuous under certain conditions. In this valley, more particularly, the streamers worked down on the back of the lodes denuded in their operations as deeply as they could go without encountering water, many rude implements for combating which have been found in modern explorations. These pits are termed "streamers' hatches," and are the foundations of many mines, but also the sources of danger and expense. Memorable instances of disaster have occurred in Porkellis by the miners working beneath or unexpectedly approaching too near these surface works, which, of course, form drain-holes for surface water, the percolation of which into the lower levels must be, in certain situations, drawn out by the aid of steam power, and by these means creating a heavy item of cost in the mines' prosecution, as is found by such mines as Wendron Consols, Porkellis, and many others, whose lodes are the result of the old men's discoveries.

In this valley are frequently found the old mills in which the ancients ground their tin ore into powder, fit for smelting, as well as the stone implements used in stamping. These are formed of porphyritic granite of the hardest nature, sometimes of a capel or lodestone, called, by reason of its extreme hardness, "Sampson." Near the present Wendron Consols Mines is one of the greatest curiosities in Cornwall; it consists of a wellbuilt underground chamber, composed of huge slabs of hewn granite. The workmanship is fine; there are places in the wall as if for lamps. Tradition says it was the place of sale for tin, but I think it far more probable that it was the residence of some old ascetic anchorite, who lived on alms, and was more R. than F; though, truth to tell, perhaps a bit of both.

At Carglaze Tin Mine, near St. Austell, the student or the antiquarian will be gratified and instructed. This work is certainly that of the ancient old men; the opening is no less than a mile round, the depth 80 to 100 feet, and there were formerly several stamping mills at the bottom. Prodigious quantities of tin ore must have been procured here in times lang syne. Even within the last 50 years it was a large tin-producing property. At present it is worked principally for china-clay, still small quantities of tin are preserved; the stratum, like that at Porkellis, is a decomposed granite; it is very white and pure. This place is considered one of the lions of Cornwall, and is well worthy a visit, either for its intrinsic beauty, as the cliffs are of fine colour and form, or as a great geological lesson, the place being full of small lodes of tin and schorls, or as, which suits my object in this place, a noble and everlasting monument of the sagacity and perseverance of the ancient "old men."

To those who visit the picturesque neighbourhood of Ashburton I would recommend a ride, or rather a walk, to Hay Tor, from which a most magnificent view of the Dartmoor tors and scenery is obtained, on the east, west, and north; on the south, the queen of watering places, Torquay, as well as Tynemouth and Newton, lie in the most exquisite panoramic positions. In descending from this eminence they may, by a slight divergence, witness the wonderfully great extent of the ancient old men's operations, as very nearly the whole of the valleys from that spot to Newton have been streamed for tin. At a mine called Bagtor these works are pre-eminently worthy of a visit; there will they witness the actual proof of the advantage taken by the moderns of the old men's discoveries. New levels have been driven at considerable cost under these old works, and fine lodes of tin ore discovered—not in an isolated case, but in several. The heaps of tin recently procured will astonish the visitors. A huge water-wheel, 60 feet in diameter, will be an easy guide to these works. Enquiry will cause direction to a particular excavation, known as the "Piskies" pits. The immense quantity of mother earth removed thence will convince the sceptical that tin to a vast amount must have been obtained. On the north part the visitor will note the great trouble and cost these old explorers must have expended to have got so deep as they have done; yet it is evident they did their best. Modern enterprise, however, seems determined to surpass the ages gone by. The gallery, or level as the miners term it, will unwater this place without the aid of steam, as at Porkellis, and I have little doubt the tin found in cutting the gallery will pay the cost of attempting this great work.

The "Piskies" of the ancient tinner were male fairies, or the warlocks of the Scotch, whose duties were to mislead folk in the night time. Many a miner who had been to fair or market has returned "Pisky laden," the Cornish term for being bewitched. No doubt he has seen more than one candle where one only was. It is supposed the word, as connected with the spot, originated from "lights on lodes," which are stated on authority beyond dispute to have been seen where large deposits of mineral are found; the light is said to resemble the *ignis fatuus*, or Jack O'Lantern. Be this as it may, the name, as applied to these works, is extremely ancient, and it may be very appropriate; if the existence of the lights on lodes be doubtful the presence of tin is certain there. The lodes may be well studied without the trouble and danger attendant on the descent and ascent of a mine. The nature of the lodes and the surrounding strata is precisely synonymous with those at the Porkellis Moor, as are also the excavations by the ancient old men. The place is well deserving attention.

The modern "old men" next deserve our notice; but as we have nearly run the length of our tether we must be as brief as possible; we feel the nature of our subject under consideration deserves such trivial remarks. Go into the mining districts and meet a modern old man, and he will give you such marvellous accounts of rich bottoms, splendid ends, and neglected courses of ore, that your very ears tingle at the sounds. On further enquiry, you learn that they are at the bottom of some shaft or mine full of water, that will cost thousands of pounds to extract before you can obtain a sight of the tin and copper ore thus foolishly abandoned when he was a boy (query). How many, deluded by the *ignis fatuus* of the myths of the modern old man, have had to pay dearly for their whistle. How many have found that the modern old man deserves the titles implied by the letters R and F? How many more will there not be who will yet be deluded by the Jack O'Lantern, and find themselves "piskey laden," if not in a "piskey pit?" The modern old man leaves not his works, but his words, to attest his deeds and his memory. The astonishing works of mining to be witnessed in Devon and Cornwall are not the results of such men, but of a noble race, who leave the mantle in the shape of ability, of industry, and of indomitable, perseverance, descending from generation to generation; yea, and shall descend so long as there be a mine to work, when they, in their turn, descend to the grave, when centuries shall have rolled over their annals; when their histories shall have been forgotten, and when, perhaps, a different race of men shall tread the scenes of their once busy life and active exertion; the student, the speculator, and the antiquarian will drop the tribute, as was done of yore—"The old men were good miners, after all." Amen.

## No Pay

"HELL TO PAY, AND PITCH HOT," WOULD BE A COARSE, but truthful expression (as the axiom is usually understood) of the state of affairs on a mine when "No pay" is announced: it is to the existence of a mine what strychnine is to the body, a sudden and fatal dose. The victim staggers, reels, and dies. The effects are precisely similar; the patient having the dose administered, is taken with a shivering, then a deadly palor o'erspreads the frame, a short struggle and all is over. We appeal to hundreds of poor captains and managers of mines if this is not a true picture, and if they have not personally witnessed the originals from which our photograph has been daguerreotyped. From the bottom of our hearts we feel for them quite as much as we do for the more immediate and needy sufferers.

The poor agent waits, hoping against hope, till the last day that a remittance may come—but no draft or order arrives: he then takes his pipe, as a solace, and comforts himself with the idea that surely the committee will not be so thoughtless as not to send the money, to disappoint so many poor hard-working men and women, who have honestly earned their pay. He then proceeds to his home, with a sorry heart and a sorry countenance, to a sleepless couch, working himself into a state of nervous excitement; he goes early to the mine, frets and fidgets himself, as well as everybody else, waits anxiously for the postman, who he verily believes never was so dilatory before. When, at last, he does appear; to the query, spoken with an assumed forced air of jauntiness, "Anything in my way to day, my boy?"— "No, Sir," is replied. The heart drops to the bottom of his belly, and the crest-fallen, terror-stricken man pauses, and asks the winds the question, "Whatever can they mean by it?" As companionship, even in misfortune, is in some degree solacing, the now half-savage, half-sorrowing man betakes himself to the clerk, if there be one, or to the second captain; after consultation, they send an express messenger to the post, to ascertain if there be not a mistake? When the answer, "No mistake," arrives they exclaim together, "Then there is no pay, and no mistake!"

Evil news is said to fly apace, and no news flies with greater or more fatal speed than "No pay;" quicker than the racer's gallop or the arrow's flight, the cry of "No pay" at G. N. P. and N. P., wings its fatal way—even echo lends its magic influence, and "No pay, no pay!" is heard on every side. The anxious merchant, at the sound of "No pay," opens his ledger to see what arrears he has in his accounts, and if it be a cost-book mine, immediately fixes in his mind's eye on the best man to take for his account, and pre-determines to be early in the field, lest he be forestalled. The shopkeeper, who in mining districts usually gives the men a month's credit for their necessaries, at once stops all accommodation, the landlord stops the tap, the poor agent, the silly-looking purser, and the poorer stricken work-people become the bye-word and taunt of their fellows. The best men leave their employ, or work sulkily and carelessly, and when complained of by the captain, they impudently answer, "What to h—l do we care; how do we know we shall ever be paid?" If on a limited liability mine such a matter occurs, the men at once obtain an injunction against the materials, and cease working, as they know they have no "good man" to take, and the game is up.

O, gentle reader, think not this picture is overdrawn. We assure you, painful as it is, it is literally true to the letter—aye, to the life. Picture to yourself, if you now can, what must be the position and feelings of persons to whom these hardworking people look up as a father and a friend (for such the captain, if he do his duty, is considered), when he, knowing they have fairly earned their wages, not only by the sweat of their brows, but by the risk of their lives— when, as on pay-days, the labourers come to take their hire, dressed in their "bit of best," in expectation of displaying that independence which often proudly marks their character, in paying off their monthly scores—when the account-house is surrounded by a waiting crowd, the poor agents, after screwing up their courage, come forward and announce "No pay to-day, my boys. I don't know how it is, but there's no money come; I suppose it will be all right after a bit." At first all is silence, then a murmur, then an out-burst of feeling, with a suppressed "D—n it." At last one of the elder men rises and asks, "When d'ye think it will be here?"—"I don't know, but I will write about it. Now, be good men; it is not our faults; there's no more money for us than for you. It is no use grumbling nor grieving; all I can say is, it is a bad job. I know that men ought to be paid, that a labourer is worthy of his hire; but what can't be cured must be endured." Then, as I have frequently known, and so late as Saturday, Nov. 1, 1862, saw an instance of:— "Any of you who are married men, and have families, come to my house, and I will let you have a sovereign each, to buy a bit of meat or fish with, as I suppose the shop will be stopped." The response is from all, even the boys and girls.—"We all have wives and families, Kappen."—"I know who is who, and shall act accordingly," is the worthy man's reply. He did, and does often, act thus; and it is acts like these that create the bond we have alluded to—father and friend; which is reality, as the poor fellow has reasonable grounds for doubt if ever he will be repaid.

In illustration of our subject we quote a few facts, the portraits and truth of which will be by many mining men recognised and admitted. At the notorious P. W. mine the men were unpaid for three months, having been kept quiet for that length of time by the assurances and little relief given by the agents, who spoke, as they thought, that all would be right in the end. Neither men nor agents have been paid to this day, though law was resorted to as a last resource. Law is a tardy remedy for starving men. The "law's delay" is a small satisfaction for angry merchants and savage shopkeepers. In the case of this mine I have seen the poor little captain dread to meet the army of despairing miserable wives of these honest, hardworking fellows, when "No pay" for a second—aye, a third—time was obliged to be uttered. I have actually in one case seen a poor little boy who knew me come and beg a penny of me to buy a bit of bread, saying "I have not tasted any for three days, Mr. Henwood." After filling the poor child, he emphatically said, "It must be a fine thing to be a gentleman, Sir." Hear that sarcasm, O ye who treat cost-sheets and poor men's claims so cavalierly as is sometimes the case, or so carelessly, as is more frequently the reason! When the cost-sheet is sent in think of this photograph, and do justice. We quote another melancholy instance of the melancholy legion. At the G. N. P. the previous scene was re-enacted, but with greater aggravation, as at that mine the men actually raised and dressed a parcel of ore. The poor men had a little subsist, but the agents "no pay" for months. To an application for a few shillings to pay samplers' fees when the ore was sent to Portreath for sampling no reply was sent: the impoverished agent was actually obliged to borrow 10s. to do that which would probably have not only damaged the mine but the sample too. What was still worse, more disgraceful, and almost incredible, is the fact that when the ore was sold (9 tons, at 7*l*. 5s. 6d. per ton) an order was sent from the committee—to the manager to receive the amount, to pay off the men, petty accounts, &c., and apply the balance to his own arrears. Not so, however; Mr. Secretary, the day before the cost was due, wrote the agent of the smelting company that he held them responsible to him as the secretary of the G. N. P., and insisted on the money being forwarded to London to him. When the poor agent applied he found the subject of our paper—"No pay." This he was obliged to say to the miners, who had been promised their money on the day. Picture to yourself, O reader, if you can, the state of this poor fellow's feelings! That poor fellow was the author of this paper.

We could multiply examples *ad infinitum*, but forbear. The task of recounting unpleasant bye-gones is at least distasteful, but passing clouds frequently yield fruitful showers; so let us trust it to be in this case. The moral of this paper is patent; let us trust its inditing may not be in vain.

And now, in conclusion, O adventurers, remember this photograph is sadly too true to nature, and is made as a *carte de visite* to you for your benefit and instruction; and you, O ye captains and agents, for your behoof; and you, O ye working miners, for your especial benefit.

### Up Hill and Down Dale

THERE IS MORE REALITY IN THE ADAPTATION OF THE ABOVE homely motto to mining practices than at first sight there would appear to be. *Imprimis*, in searching for mines it is literally correct, "Up hill and down dale" being almost only the conditions under which pioneers and explorers can obtain experience and the object of their pursuits. Feather-bed miners and stay-at-home travellers can never expect to be successful, or to have confidence reposed in them; the persevering, hard-working man is the man to whom the mining adventurer should look up when seeking an investment. Were such the case, disappointment would be far less frequent, and mining *per se*, as well as its workers, spared undeserved obliquy and illiberality frequently administered by ignorance and presumption.

The subject of our paper may be, and certainly is, a very interesting pursuit, so far as this sectional view is studied, but it will be found in practice to be fatiguing, and not unfrequently dangerous employment, always expensive; these considerations should be borne in mind when persons about to enter on mining read of what appears to them vast sums for preliminary expenses. Only let them try the experiment, and they will see and feel the truth of our remarks.

Another view of our photograph is afforded by what must be acknowledged the chances and risks of mining adventure, for such they are after all. Were they all prizes and no blanks, mining would cease to be a speculative undertaking; the business would cease as a science; everybody would be miners—it would be worse than now, when there are already a host of professors. There always has been, and always will be, ups and downs in mining interests as well as in other legitimate businesses, but who has not witnessed the "Up hills and down dales" of mining. The history of nine mines out of ten attests the fact and the illustration. How many hundreds of cases could be cited in which once prosperous mines have met with adverse circumstances, and have yet again been resuscitated and restored. The miner's motto should be *Nil desperandum*; without that idea and determination his life would be, indeed, a hopeless one. Were he to be cast down by one stroke of misfortune or bad luck, he would be wholly unfit for such an alternating business. It is a long lane that has no turning, and it would be long and dreary, too, if it had no end. We are proud, however, to say that the true, genuine miner, who knows his business, throws fears like these to the winds. He pursues the even tenour of his way, in full confidence that if there be a cloud at one time there will as certainly be sunshine at another; that the adventurers' risks, like the mine's history, will resume its wonted good time coming, even if they have to wait a little longer. If "Up hill and down dale" be applicable in the former instances, how much more is it to the case of the poor working miner, who plies his toil at such depths from the surface. The toilsome ascent and descent to reach their work is, indeed, a true and natural subject from which our photograph has been taken; theirs is truly a life of hard work, yet they seldom complain. They have their ups and downs literally in more senses than one, especially the tributers; who may be poor to-day and comparatively rich tomorrow. We have known many instances in which working men have made from 100*l*. to 200*l*. in a month by thus speculating; in others, where the same men had worked for two or three months and not earned a shilling. How many mine agents and captains can vouch for the accuracy of our motto. How many of them have we not seen on the pinnacle of success, at the head of their profession, and luxuriating in the sunshine of prosperity, who in a brief period, by a stroke of adversity, have been reduced to comparative penury. It would not do for them to fret or funk. No, no. They take heart in misfortune at the remembrance of the subject of our paper, and as certainly rise again.

The practical purpose of our essay is to warn persons not to prejudge any person or thing connected with mining because an untoward time may occur. We hope we have stated enough to inspire confidence; we could quote examples *ad infinitum*. Our immediate object, however, is to quell the fears, dispel the prejudices, and inspire the hopes of certain parties with whom the author has been concerned. We need not further allude to them, for obvious reasons; we only advise them to take proper means, employ experienced exploring agents and experienced pioneers, and not be guided by or listen to faint-hearted, ignorant, and presumptuous professors, whose only knowledge consists in finding fault, and taking a negative in everything, when all will be right. It would be as unreasonable to expect such parties to manage mines, as it would be for an uneducated Cornish miner to edit "Frazer's Magazine" or "Buchanan's Sermons." If would-be professors would let mining matters be to mining men, we should have far less dictation and nonsensical experiments, and certainly far less "Up hill and down dale" in mining interests. We hope never to witness the day, however, when the practical utility of the subject, as applied to exploration, shall be neglected, but to see "Up hill and down dale" preeminent; to see youth practising the wholesome, invigorating exercise for their own benefit as well as that of the distinguished professions; by these means will they become to it what it sadly wants—not a dictatorial humbug, but a *Decus et tutamen.*

## The Mining Pioneer

It is surprising to witness the variety and difference of characters engaged in pursuit of this difficult and dangerous avocation; yet each, when properly imbued with the cacoethes, following it with an ardour and interest amounting, in many instances, to a kind of monomania or frenzy. Yet how few, in this country at least, meet with a due reward for their exertions, still all endeavouring to arrive at the same goal—wealth and fame. I cannot at this time, nor ought I to, omit the name and memory of a distinguished mining pioneer, one who neither spared time, wealth, or trouble in hunting the hills, dales, river-beds, and other spots likely to afford success for indulging his favourite hobby—for such it was to him. To him we owe discoveries, the importance of which, in reality, we have not yet begun to appreciate. Doubtless, ere long, and when railways shall have penetrated into the wilds and recesses of Perth and Argyleshire, the value of his labours will be felt by a labouring population, and by all the beneficial consequences mining occupations convey in their train.

The late Lord Breadalbane was, indeed, a remarkable instance of perseverance to this particular branch of science. He not only took a delight in roaming over the hills and dales of his vast domain, with pick and gad—which, by the way, he knew how to use effectually—but he courted and enjoyed the companionship and conversation of scientific and practical men. He might have been said to have been the Macenas of mining. By these means he arrived at no mean or trifling knowledge of the science as a practical as well as theoretical miner, his wealth and position always giving him unusual opportunities of inspecting the rarest collections of geological and mineralogical specimens, as well as visiting their habitats, and viewing them *in situ*—a wonderful advantage to an ardent student, as was he, notwithstanding his dignified social position. Oft have we witnessed the scenes of his labour, and oft have we seen his untiring energy in searching the most remote and almost inaccessible places for their metallic produce. We fear we shall not look upon his like again. Peace be to his ashes.

Come we now to a much humbler individual, but one no less useful in his sphere of life, one equally energetic and persevering as the greatest, and one whose practical utility was not only appreciated, but, in some degree, met its reward. How seldom is it that humble and modest worth meet these desiderata. Poor Dick Stephens, perhaps better known by his sobriquet "Paddle your own canoe"—an apothegm admirably adapted to mining, of which more anon. Well known was he in the neighbourhood of the Fowey Consols; oft have we journeyed with him from early dawn to dewy eve, tracing cross-courses, elvans, and lodes; to him time and money were but of little account had he an object in view; to him the rocks and stones were but as household words, the lodes as the alphabet, and the mode of working them as but his first lesson. He was deemed not only a first-class miner, but a first-class mining ferret; yet he had his eccentricities, but, as our motto is *de mortuiis nil visi bonum*, we pass them over in silence; they possibly cost him his life. So highly were his merits held in esteem that many of the Cornish land proprietors gave him a *carte blanche* to go anywhere and examine any part of their estates, without let or hindrance. Cornwall presents a vast and strange anomaly to most parts of Great Britain; in that county, where mining and its value are known, miners are courted, and search for mineral encouraged by kind attentions and liberal dues, instead of frigid doubts, grasping dues, impossible conditions and restrictions, hasty demeanour, and ignorant pretensions, as in many parts of Scotland and elsewhere, in places where mining is unknown, or but little practised. We introduce this digression for obvious reasons—contrast gives double effect. This unlimited license, as might have been reasonably expected, led to a thorough examination of the estates, far more efficient and effectual than if the proprietor had employed the whole batch of professors and pretenders; aye, at how much less cost? One of these worthies, the Member for East Cornwall, was so much pleased, and so well satisfied, that he would not grant a sett over which Dick had been for discovery except the applicants made him adequate remuneration; this was a *sine qua non* well known to many of my readers. The subject of this notice was accidentally killed on returning from one of his excursions.

Another pioneer, whose name deserves record as a most remarkable instance, was Robert Pease, of Hawes, Yorkshire. He was almost a counterpart of poor Dick Stephens, alike almost illiterate, save self-education; like him the lark and the crow witnessed his outgoings and home-comings, and like them, the field and the wild moor were the scenes of his pleasures and his labours. To him the hill-side, that appeared only fit for a sheep-walk or an eagle's eyries, were as conversant as Oxford-street to a Cockney; to him the ravine and the waterfall were as well known as the Thames to a waterman, or as the Bank to the capitalist. To him do we owe many not only useful hints but absolute experiences in the rocks constituting the great Yorkshire mining district for lead and copper ores; the latter not half-tried, and but barely examined. To this pioneer do many mining companies owe the information that engendered their very existence; his knowledge was great, though his patronage was small, and that little shabby. He had no Cornish worthy to take him by the hand; he had those who not only stripped him of his knowledge, but would have stripped him of his skin could they have done so, poor fellow. He paid the debt of Nature with the coin he brought into the world. He had not wherewith, except Mother Earth, to lay his head. He allowed others, but did not "Paddle his own canoe." The quaint maxim is very suitable, and should be acted on by every mining pioneer—they should always paddle their own canoe. If they make discoveries make the best use of them. They may rely on it that few Mr. Kendalls are to be found. To such men as the noble Marquis, first quoted, reward would be but of little moment, yet even he would not give away his discoveries unless adequate consideration were paid in the shape of dues. He certainly was liberal in his outlay, but he would not allow advantage to be taken. He knew how and adopted the apothegm. Stephens did so to a degree, but in the instance of the poor Yorkshire pioneer, who was simple enough to tell all he knew, and display his discoveries, when he wanted to beg or borrow nobody had a paddle or a penny to lend.

The principal end and aim of our essay is to demonstrate from the life, by true pictures, that the profession is a worthy one; that it has been, and is, practised by all classes, from the peer to the peasant. That it frequently, but not on all occasions, meets with a suitable reward; that consummation is to be realised partially by the party's own management and conduct. That the benefits derived are by no means selfish, even though a suitable reward be claimed, I think has been shown and demonstrated in previous essays—is, wherever the footsteps of the miner tread wealth, and general prosperity assuredly follow, bringing happiness, contentment, and civilisation as companions. But, O pioneers! let me implore you to observe poor Dick Stephens's motto—"Paddle your own canoe."

## PERSEVERANCE AND SELF-CONFIDENCE

"PERSEVERANCE" AND "SELF-CONFIDENCE" MAY BE JUSTLY deemed the mainsprings of mining adventure: without these elementary principles the most lavish subscriptions would be futile, the richest mines inert, and the speculations ruinous. They are to a mine what the heart is to the body, the sun to the day, and the stars to the night; without them all would be blank. With them success, though it cannot always be commanded, is yet rendered next to a certainty. Without self-confidence no man is justified in attempting an outlay. Having once fixed that stirling virtue in his mind, "cursed is he who, having put his hand to the plough, looketh back," is the most applicable proverb we can recollect to express our sentiments. How many disastrous cases of mining speculation can we of ourselves recount, in which, from the lack of one or both of these principles, mining has been condemned wholesale as a ruinous, gambling, lying profession? "Aye, profession! and nothing but profession, without the shadow of a reality," methinks I hear some disappointed adventurer angrily assert, without thinking that he has most likely been the cause of many disasters, he wanting the principles above quoted. He most likely has never tasted the sweets of reality; and, having suffered disappointment in one case, his temper is ruffled, his ardour cooled, and his spirits depressed; whereas, had he the support of our text, he would as assuredly ultimately triumph. We are involuntarily led to this lengthened proem by a circumstance which lately came under our notice: for fear it might be deemed a puff, we shall omit names. To many of our readers the original of the photograph will be well known, and I doubt not recognised; at the same time, it bears so great a family likeness to the whole of the mining generation, that it may be taken as a picture of mining in general. That it is not an isolated case will be accorded even by the most bitterly disappointed and savage speculator, when he growls out the expression before mentioned in his anger. Suffice it for us to say the good fortune has fallen on a man every way deserving, possessing, as he does, these virtues in an eminent degree.

Being in Ireland some five years since, inspecting mines, I was invited by the worthy to whom I refer to visit his mine. "I fear we are at the far end," quoth he. "I should like to have your opinion; our party have lost all confidence. We had a little ore shallow; we are now down to the 35 fm. level, and the mine is poor. Prettier indications you never saw, but our capital is exhausted." "Well, Captain," replied I, "the old tale; let us see it." On visiting the mine I found a very good plant of machinery, the whole surface arrangements well and carefully laid out, a burrow from the levels and shafts, and such a burrow! I exclaimed, "What! are the proprietors mad to suspend and ruin a mine in such a state and position as this? They ought to be hanged, drawn, and quartered. Why they are

about to leave off just where they ought to begin? Do they expect
to find bundles of ore amongst the fern roots, or do they expect the
ore to raise and dress itself at word of command? The appearances
here indicate to me that you will have a deep and lasting mine."
"That," replied he, "has always been my opinion; I am glad to have
it confirmed by you." This visit tended, for the time, to restore
confidence; the mine was not abandoned, but was gradually
deepened; still the promising reward was not forthcoming. The old
despondency once more became ascendant and overpowering. At
length the ore was found in the 50; but, as it could not be raised *ad
infinitum*, and converted into cash on the instant, so as to pay
dividends, the discovery was deemed by the savage adventurers as
less than nothing. "Better be kept in longing expectation of something worth while than reap an empty reality," was their cry; and
"stop her, stop her," their command. The captain, having perseverance and confidence on his side, begged them to hold a little
longer, to wait the event of the other 10 fms. The more violent
insisted on not expending another farthing. The mine was on the
eve of "knocking," when it was suggested by the agent, "Why not
sell out part if not all your shares? But take my advice and hold."
"Sell out! farce, nonsense; who will buy? Who will be mad enough
to buy in a mine that has been fairly tried for seven years, and nearly
ruined everybody? Nonsense; who will buy?" "I will", replied the
captain. Business was transacted to mutual satisfaction. This effort
for the time again procrastinated the mine's suspension; the works
proceeded. As the mine got deeper the lode gradually improved,
but not in the manner the impetuous shareholders required. One
after the other they expressed their disappointment and disgust,
abused the whole affair, and all such in no measured terms. The
captain, still relying on his principles, worked on, and on. In the 60
and 70 fm. levels he and those who abided by his advice and experience met their reward. In the bottom of the mine they have now
the finest course of lead in Ireland, and the mine is valued at more
than ten times the outlay. The value of the shares at present may be
counted by pounds, instead of shillings as they were purchased
from the malcontents, to their infinite chagrin and disappointment.
The captain, as a matter of course, is by them branded as a scheming
villain, who knew what was there, but kept it secret from them!
He should not have allowed them to have disposed of their interest,
particularly to himself. Those who stuck to him and his creed look
on him as an oracle, and so do his faulters, if we could see their hearts.

Had this miner not been largely gifted with the noble qualities forming the subject of our paper, this source of wealth, of employment, of happiness, would probably have lain neglected for this generation at all events. Now, it is calculated to be a real blessing, not only to the proprietors, but to the whole neighbourhood. All the ground in the locality for miles has been secured for mining pursuits. The nonsense of the ore existing in shallow deposits only has been exploded by the facts; and the once busy neighbourhood— for the ground has been worked in days lang syne—will resume its ancient activity, to the comfort and well-being of hundreds, as well as to demonstrate Ireland's mineral resources.

In this instance, as in all others, the excitement caused by discovery is apt to outrun reason. However, the proprietary being limited to a small number of individuals, less of it has been felt, and the facts less known, than would have been the case had the mine been in the London market. Miners have been busy tracing off the lode into other setts. Operations have been commenced with surprising results, the which will ere long be made public by the sales of produce, the best test. All this great and good work and achievements have been brought about by the perseverance and self-confidence of one individual. Go on, worthy man, thus; may the same fortune attend you and all such! When such principles lead the way, Fortune will be sure to walk in the path. We have before said this is by no means an isolated case, nor is it so far as our subject is concerned; but it seldom happens they who deserve meet the reward, as in this instance.

Now for the converse or, rather, reverse of the picture. How many mines have been irretrievably ruined, not only the mine themselves, but the adventurers,—the whole locality condemned and tabooed; until, at a future period, some such able-minded personage takes the matter in hand? How many? Echo answers—How many? The Irish echo would reply—Hundreds upon hundreds. Within our ken during the last month we could name half-a-dozen most deserving properties suspended because pig-headed, stupid, vacillating, ignorant, greedy committees and adventurers expect the mines to produce shiploads of copper at the surface, that shall produce 20 per cent. for copper without dressing! One mine, producing over 1000 tons above the adit, is to be suspended: as the captain cannot find the lode in the 10 fm. level, after a few weeks' trial, the company will not risk the capital to sink the mine. In another, the mine is suspended to ascertain how its neighbour fares? Yes, so it is! I have actually been asked, in reference to a lead mine just opened from the surface, in which the men had been at work a fortnight—"How soon do you think they will ship a cargo of 30 tons of lead?" Hear that, O ye miners! and hear that, O ye captains and pursers! would your ears not tingle at the sounds?

As we do not believe in tales being useful without morals, we apply our idea of the benefit to be derived from what we have written—That perseverance and self-confidence are the dicta which should be the miner's encouragement, that they should not be cowed or discouraged by the fears or threats of pompous, over-bearing chairmen or committees, and foolish, silly, cow-hearted adventurers, provided their mines be in good situations, and present indications which they from analogy deem worthy a trial; not to despair if they cannot send off a cargo of 30 tons in the first fort-night's working, or if they do not cut the lode in the 10 fm. level, when carried off by a slide, after selling 1000 tons from above the adit.

And, adventurers, a word for you. Do not put your faith in com-mittees who have no knowledge, and seek an abundance of opin-ions, having none of their own; but if you have a captain who has perseverance and confidence to purchase your refusals, and who imitates the worthy man who has been so truthfully represented, emulate the example of them who followed him, and you shall assuredly reap the reward always attendant on Perseverance and Confidence.

## The Mine Sale

"THE MINE SALE"—THE *finale* OF A MINE ADVENTURE IS not unimportant, but is an imposing affair, if of any considerable magnitude. Both these terms may be applied in double *entendre*, each application being equally apposite. Important it undoubtedly is to Mr. ——, auctioneer, and the bargain-hunters usually attendant on such occasions; important, in truth indeed, is it to the disappointed shareholders, yet in how different degrees to each so concerned! To the former, a windfall and a benefit; to the latter, a fall indeed—a blast of their fondest hopes and anticipations; realising to the life the apothegm and sadly o'er true tale, that what is one man's food is another man's poison. Imposing is it to see the array of business and no-business bodies who attend such scenes, the majority to prey upon the wreck which law, too frequently driven by ignorance, has created. If the fangs of its myrmidons be once placed on the victim little chance of escape remains. After battening to repletion, the refuse and remains are handed over to harpies, who devour and destroy without compunction. To all who have ever been to a sale of materials on a discontinued mine the application of the proem above recited in its double sense must be self-evident. To those who have not had that opportunity we propose a Photograph, which we trust will be to them what the panorama is to the tarry-at-home traveller—to ensure a faithful representation of the scene, without the discomfort of a wet jacket; for, gentle reader, be it known it rains at a mine sale as truly as it does so nine days out of twelve in Manchester or Pontypool.

No sooner is the judicial fiat pronounced, the commission for the sale issued, and the order given to good Mr. ——, under "certain restrictions and conditions," known only to legal gentlemen and Mr. ——, than this worthy betakes himself to his office or study, to George Robin-ise the property. No compliments to his employer, or glowing descriptions, are deemed too lavish to express his estimate of the favour conferred on him, or of his high opinion of the value of the property. This is sometimes carried to such excess as to excite the merriment of those who understand the real state of the matter. One condition is seldom omitted—"Refreshments will be provided at the commencement of the sale, and at the conclusion for purchasers." This is a most important and imposing item in the showy advertising handbill forthwith issued, as it is far more certain to secure a numerous attendance than the intrinsic value of purchases to the majority of purchasers. Mr. —— feels and expresses himself in grandiloquent terms how highly he is honoured on the occasion, which honour he displays in the shape of consequence and bustle—a pardonable frailty. Turkey cock-like he struts about, and counts up the probable gross assets of the sale—ergo, his gains—whilst the poor adventurer stalks like a ghoul, or, like Volney, sits and studies the wreck and ruin of empires, or of what was of paramount importance to him. There lies the debris—the remains of what so lately was proudly looked upon as valuable property—drawn from their dark recesses, dislocated, broken, mutilated, or hurled from their proud positions, exposed and ignominiously treated. Yes, these tremendously costly items and specimens of engineering ability that had, perhaps for years, made such frightful costly particulars of merchants' supplies in the monthly cost-sheets, now made palpably evident to the sight of where all the enormous supplies have been employed. There they lie, confused masses of pumps, bobs, H-pieces, glands, strapping plates, match pieces, old timber, scrap-iron, and all the thousand-and-one things usually discovered on all occasions when a regular turn out takes place, so that it pretty well resembles a sacked dockyard. The number and variety of materials fairly stagger the adventurer, who involuntarily ejaculates—"Well, after all I did not think there was half the stuff here; I can now perceive, in some degree, how the money goes in mining undertakings."

The day of sale having arrived, a motley group gradually assembles, of all grades and conditions in society, from the gay post-chaises of the auctioneer and law agent, the burly-faced mine captain, the bargain hunters, and resident farmers, down to the hobnailed boor, the unemployed miners, or sturdy beggar boys, who go, they say, to see what is going on, but in reality to get a "drop of summat," as they are sure to read the magic words "refreshments provided," and hope to share in the spoil. This is sometimes carried out to a shameful extent, but, of course, is put down to the cost of the mine sale. After waiting an unduly long time, the auctioneer appears, hems two or three times, and delivers his exodium—the conditions of sale,—which generally elicits some grumbling remarks, one party desiring the whole to be sold in one lot, another fully understanding the sale was to have been in detail, and others wanting only particular portions. This over, the tables are turned on Mr. Auctioneer, everything is found fault with, or is valueless, for no measured terms are used to depreciate the property. There invariably happens to be a favourite or an auctioneer's agent present, who gets, or "picks up," the best bargains at a nominal price. In a short time the same articles may be seen in the broker's yard at cent. per cent., and something more, advance. As the property is seldom required in the locality, there is usually a tremendous sacrifice on such sales, in addition to the charges otherwise laid on when a company of adventurers have to pay the piper; it is, therefore, no wonder the assets are generally so trifling and disappointing. Many are the merry jokes and boisterous laughs at some would-be witty sally of Mr. —— —; in the meantime the refreshments go merrily round, little heed being had to expenses.

Ninety-nine out of a hundred of these cases arise not from the poverty of the mines, still less from their exhaustion, but from the wilful obstinacy of some grumbling, discontented shareholder, who because he cannot have his own way precipitates the adventure into ruin. He thinks it is only to run to the lawyer, get him to meddle, and he will be enabled to be the cock-of-the-walk, and dictate. We could enumerate many hundreds of instances to the point, in which the small cloud has grown from the size of a man's hand to an overwhelming storm, such storm sweeping all before it, even to the scenes we have described as the mine sale. We have known one in which the neglect of payment of 8*l.*, arrears of wages, led to writs being issued to sundry adventurers, who, imagining themselves to be aggrieved parties, plunged the affair into a Chancery suit, whence it came not out until the mine sale took place, the property that cost thousands sacrificed for an old song. The adventurer, and adventurers too, were overwhelmed in one common fate of ruin, after endeavouring to extricate themselves and their promising mine from impending ruin. Had the obstreperous shareholders possessed common sense and common honesty, instead of throwing their property away in such folly, endeavoured to agree instead of quarreling, they would have realised and enjoyed the prize which has fallen into other hands.

In this paper our purpose is not to condemn or prejudice mining— far otherwise; it is to caution adventurers against precipitate law actions, and discontented committees; for, as certainly as discord enters their councils, it is ten to one they witness the destruction we have endeavoured to describe; that they will subject themselves to the prying of the harpies, from the principal officer to the hobnailed boor, who will laugh in their sleeves whilst they quaff the refreshment provided at the cost of the mine; and that they will have the melancholy mortification, when too late, to be as Volney or Marius, and wonder at the folly and proneness of their fellows.

We know that under the limited liability mode of working mining companies the risk of incurring debts which nobody is willing to pay is far less imminent than under the old Cost-book System; still we know, from painful experience, many causes arise, even under the new and improved, *regime*, by which mining speculations are, and will be, brought to ruin, to which end nothing can be more conducive than dictatorial chairmen, ignorant committees, and grasping directors. We implore all who may be subject to such annoyances to endeavour to meet them by temperate timely advice, such as in the end and aim of our essay, when they will see fewer sacrifices in the shape of mine sales, and we trust will never have cause to lament over the subject of our Photograph, which may be depended on as being faithful in every particular, in cause and effect; and we appeal to hundreds of mining men if they have not witnessed many parallels in their experiences.


## A Leary Belly Makes a Saucy Tongue

Is one of the truths so unpalatable to the parties who are the cause, as well as to those who are the passive utterers of the maxim. It is an easy matter to judge on the first visit to a mine whether the men are regularly and sufficiently paid. In the one case a smart activity and readiness to answer questions, and an obedience to the worth or dread of the threat of the captain at once indicates good pay and good management; but where a louting gait, with hands in the pockets; the miners lolling against the smith's shop doors, with pipes in their mouths; when addressed by the captain, a short, cutting response bespeaks, "post pay, if any, on pay day;" they are as certain indications as that the presence of lightning will cause thunder, or that the advent of an adventurer will cause grumbling, starving men and pining wives and children to complain.

On being remonstrated with by a visitor for their heedless appearance, and evident unwillingness to work, the answer invariably is, "What to ——— is the use of our working when we gets no pay. Kappen, poor fellow, es no better off than we poor fellows; he can't help it we know, but 'tis no use to tell the shop-keepers so; they wont trust es but a month, and then what is us to do? We don't want to go to court about it, but the shop-people do put us there too; so if we are not paid we can't work, and you do know our work is hard, and requires good meat, and plenty of it, which we seldom gets. We are now two months behind, and from what we can see there's no pay for this. I only wish to God I had the money, I'd soon be off to Australia."

Underground the scene is the same, only more gloomy. This state of things is a grievous error, and should by all means be prevented. No wages on pay-day is to a mine what a returned bill is to a merchant, or "no effects" is to a gent keeping a bank account, on the presentation of his cheque in part payment of his tailor's bill; all confidence is lost. The exposure is not half so bad as the after consequences. The poor men go home to a dismal hearth; discontent reigns paramount; their poor wives wreak their disappointments on their innocent offsprings, and the old couplet, by a slight alteration, may be faithfully realised:—

| | |
|---|---|
| The very kittens on the hearth, | The very devil's in the house, |
| They dare not skip and play; | If I don't bring home my pay. |

There can be no doubt of the social evils arising from these causes being far more extensive and ramified than is usually considered. Little does the angry though rich shareholder suppose that his not sending his cheque for a few days after its becoming due may lead one of these hard-working men to a prison, for trapping a hare or rabbit, stealing a few turnips, or compelling him to plunge himself into inextricable debt, for which accommodation he pays, or rather is charged, double price, in the latter case causing domestic affliction incalculable; for when a man recovers good employment his wages are so scanty that he cannot recover his status, being ground down by gross overcharges, and the cursed laws of this country, that if a man once becomes poor he must for ever remain so, by the peculations of her numerous myrmidons. In the former by compelling a man to thieve, merely to satisfy the calls of his famished children for nourishment, to quiet the reflections of his angry, half-distracted, wretched wife, or to renovate his system, fatigued and broken by the most severe and trying as well as dangerous labour known to man, entailing misery to them all, expense to his country (for to prison he must go), and a disgrace indelible to those who were the cause. Think not, say not, there is the workhouse, which will give a little relief at such a time. If you think so, dispel the vision; a rough rebuff and insult would be the reply to the starving man, woman, and children. Go to the law for your redress; and whilst this tardy process is being carried out to gain their rights, their necessities prevail over their better judgment, and they are for ever lost. Think not we state what is untrue; but oh! tardy adventurers, pray remember our Mining Maxim, and rest assured it is true, sadly too true, that "A hungry stomach makes a saucy tongue."

## "THEY THAT CAN'T SCHEMEY MUST LOUSTER"

TEACHING BY PROVERBS SEEMS TO HAVE BEEN PREVALENT IN every age and in every state of society. Of the former we have abundant proof in sacred and profane history. Amongst the most savage, as well as semi-barbarous, nations the same method is found to exist. They seem intended as a series of brief homilies for committal to memory, and are certainly more likely to be remembered than more extensive or lengthened wise sayings.

Perhaps no axiom is better suited or calculated to encourage miners in their avocation than the proverb at the head of our paper. Its truth is equally apparent as that forming the subject of our previous paper. Ingenuity and contrivance are qualities for which the Cornish miner is pre-eminently distinguished. An instance has just come under our notice which fully displays the truth of our assertion. A man named T. Sandercock, who had for years been accustomed to work in deep mines, was determined, if possible, to render himself capable of taking charge of a steam-engine, and securing a good situation ere declining years came on. With this view, and with the most limited means, he set to work in his unemployed hours, and built a model engine, with a horizontal cylinder of $2\frac{1}{2}$ inches diameter and 7-inch stroke. This author of the paper was accidentally passing the cottage when the first trial of this ingenious and creditable piece of machinery was being made. The extreme satisfaction of the poor fellow on finding it worked to admiration was only equalled by that of the two or three gentlemen who were present, who also rejoiced as he rose up with pride on his countenance, and gratification in his soul, and exclaimed, "Surely I shall be able to get a place as engineman now!" does not this man deserve encouragement? We hope this paper may be the means of forwarding it, as he bears a most excellent character for sobriety and industry, of which no better evidence than the foregoing could possibly be adduced.

If the miner cannot plan his work cleverly he must work hard or starve. Necessity is the mother of Invention: this principle, a taste for and readiness to adopt improvements not often found in workmen of the grade to which miners are supposed to belong. On some occasions the introduction of what they term a "new fangled notion" causes a little opposition and obstinacy (as in the case of the safety-fuse), when the captains say, "Well, they that can't schemey must louster," objections are silenced, and the trial made, far more readily than if the most elaborate and persuasive language had been adopted. This phrase has also been the cause of many a fine and intelligent fellow raising himself from the rank of workman to that of superiority. By contriving and endeavouring to excel he so far improves his mind and life as to render himself independent of the necessity of drudgery.

Like all others, it too has been abused, and many have they been who have taken to scheming in its worst sense, by which to avoid hard work. This, however, does not militate against the utility of the apothegm. There is scarcely one in the whole list of proverbs that may not be twisted to suit some vile end or witty contradiction. To those who endeavour to accomplish such ends we leave the merit and the profit; we look only on the words as they are obviously intended to be understood, and regret that so many of our old Cornish mottoes, proverbs, and nearly obsolete but expressive and significant words, should be allowed to sink into oblivion. They would, doubtless, form an interesting volume, highly valuable at some future day, as many an old story, and indeed documents of extreme consequence, abound with such quotations.

We commend the subject of our present paper to the notice of miners generally, but to the younger portion in particular, and urge them to its consideration in the point of view we hold it. By habits of attention, observation, by continual experiment and contrivance, to prevent the necessity of extreme labour, that by so doing they may raise themselves, save mining expenses, and so encourage and stimulate mining adventure; and we beg them to remember the example of Thomas Sandercock, the miner, and the boy who by "scheming" discovered the method of self-acting gear for opening and shutting the valves of steam-engines, by which he saved himself from "louster;" that the field of invention and contrivance is as yet but just on the dawn of cultivation by mankind; that it is illimitable, and that all in their sphere have it in their power to practically improve the old Cornish wise saw—"They that can't schemey must louster."

## "IT DED FOR FAYTHER, AND 'TWILL DO FOR ME"

HOW RIFE AMONGST MINERS IS THIS MAXIM! STILL, IT IS not wholly confined to this peculiar class, though by them frequently quoted, and, we regret to say, still more frequently practised. Until within the last few years, it was next to impossible to thwart the proverb, or to convince the miner of the benefits resulting from improvements. A great deal of the old leaven still exists—the phrase is far from obsolete.

We have been induced to dilate on this subject in consequence of a recent visit to a great and celebrated mine, where we found, to our astonishment, it was in full activity. We suggested an improvement we had witnessed on another equally famous mine in their method of tin dressing. The person showing us over the mine admitted the superiority of the new mode, and the imperfection of theirs; when out came the maxim, or its full equivalent—"Our owners say it did for years before, and it will do well enough still. We pay dividends, and they are satisfied; they will not lay out a pound in trying experiments." This paper will be seen by the party who made the observation; he will recollect the circumstance, as at the time we said we should endeavour to improve the occasion. We hope he will direct the attention of his adventurers to it.

Now, nothing can possibly be more injurious to the true and substantial mining interests than the practice of this silly system. British mining must progress and keep pace with the times. In the present state of society, they who hesitate are lost. If adventurers in mines intend to retain their properties, and maintain their profits, they must discard all such notions; if they do not, others abroad most decidedly will, to the serious loss of the interests of all classes in this country. We could mention many mines where the practice of the adage had precluded advantages to such a degree as to necessitate their suspension; the old fashioned methods of working and dressing the produce rendering them wholly unremunerative. The advocates of progress then stepped in and showed them, painfully and practically, their egregious error and folly. We could point to a mine in which this obstinacy and heedlessness had well nigh proved ruinous to all concerned; though the adventurers had been implored over and over again to adopt recent discoveries and facilities; yet no. Their cry always was, it is too expensive to try; it has done hitherto, and will do still! Now, however, the mine has nearly changed hands, through being sold from hand to hand, until the price became literally nominal, the purchasers have determined to erect the requisite machinery, with every probability of shortly having a dividend mine. The ore will be returned at one-tenth the previous cost, will fetch a better price, and hundreds of tons of what was called refuse be sold at a profit.

Had the Willyams, Taylors, and such like gentlemen, acted on the principle that forms the subject of our paper, where would their gigantic fortunes have been? Echo answers—Where? Where would the employment of the persons who occupy so many happy homes have been found? Where the thriving villages of which their names form the nuclei? Echo repeats again—Where? Not only do they adopt every important improvement, wherever to be found, or by whomsoever discovered, but they encourage and reward their ingenuity by patronage and support.

The chief object and effort of that admirable institution, the Royal Cornwall Polytechnic Society, and of the Mining School of London, have been to eradicate the remnants of this antiquated but foolish saying, and to raise in its stead the motto of man's best practice—"Onward, onward!" This should be the practical and practised cry of miners, as well as of manufacturers. Were they as slow at adaptation. Manchester would still have been but a village, Liverpool a collection of fishing huts.

We do not mean to charge all with this apathy. We know there are many brilliant examples amongst miners, who do all they can, and labour hard in the good cause, both by precept and example: nevertheless, we cannot disguise from ourselves, though we may lament the fact, that there are yet a large number of pig-headed, obstinate, ignorant people, who will not be convinced, to their own benefit, but who still cling with fond affection to—"It ded for fayther, and 'twill do for me."

## SHALLOW MINES, AND DEEP CAPTAINS

MUCH MIGHT BE WRITTEN AND SAID ON THIS SUBJECT, which at first excites a smile on the reader's countenance; it, however, like most old wise saws, cuts two ways, according to the manner in which it is applied, and both presenting equally truthful adaptations,—in the one giving a faithful and deserved compliment; in the other, conveying a biting, bitter sarcasm. As we expect few will apply the latter to themselves, we shall offend none. To Cornishmen, and all connected with them, the truth of the adage is apparent; we, therefore, leave its application to the choice and discrimination of persons, not doubting they will, in their experience, have found some suitable subject.

The writer of a letter which recently appeared in the *Mining Journal*, on the subject of reworking old and deep mines, seems to have had this old and practical mining maxim in view, for he warns your readers against scraping out the egg-shells of deep mines, and advocates the working of shallow mines by deep [1] [2] captains, which, if carefully managed, are almost sure to yield a profit in the aggregate to adventurers, whilst the prosecution of the contrary too frequently only gives a profit to "deep captains." These may be partially true, and we admit them to be two applications to which the motto may be adapted; but these are not the only cases. Do we not see many "deep captains" employed in other professions than mining? If we ask ourselves that question, we see the universality of its suitableness; and therefore, though couched in and applied to the Cornish more particularly, it is true to the letter; hence we say and argue that it is a pity many of our old sayings are becoming obsolete. Well would it be for all if, when thinking over these brief homilies, we would apply them (as they are intended) to our advantage. How many hours of heartache, of bitter regret, would the study of the *double entendre* at the head of our paper have prevented?—aye, how many! We hope it is not now out of place to recur to them, and to beg our readers to well consider the motto; for it may be depended on as literally true, that if they enter on the business of mining no book that ever was written on the subject could afford better advice or more wholesome warning. We do not mean to infer that deep mines should not be wrought—far from it; but let them be wrought gradually and with sufficient capital; let them be confided to experienced deep captains and miners, not to novitiates, who, from want of practice, have want of faith and perseverance. But to all we commend the witty old apothegm—"Shallow Mines, and Deep Captains."

## A CLOMEN CAT IS POOR HAVAGE

EXCEPT IT BE EXPLAINED, NO ONE BUT A THOROUGH WEST country Cornishman can really and thoroughly comprehend or estimate the truth and pith of this axiom. The metaphor of the "Clomen cat" is completely Cornish, as may be seen by the word: it means, hollow to the toes, or every inch deceptive—these images being, of course, always hollow to the toes. We question if in the range of proverbs there is a better simile, when properly explained, than this homely and truly western saying. As we have before observed, many of these trite saws are of ancient date; this un-

doubtedly is so though becoming obsolete, none of its moral force
or adaptation is diminished. Long may such homilies exist amongst
our mining population. Independence is their pride, degradation of
any kind their horror. The workhouse is an abomination to them.
If any have relatives in distress they strain every nerve, exert every
energy, and suffer every privation to keep them from a fate that
would at some future period be thrown as a taunt on their fair fame.
With these hard-working people there is a certain pride of ancestry,
a boast of descent, and a veneration for heir-looms that can hardly
be credited by those who know not their idiosyncracy, and have not
studied their peculiarity of character. More is to be gleaned by the
care bestowed on some piece of furniture, or old garment, that
"was worn or used by my good old grandfather—people were
better then than they are now" than from storied urn, animated
bust, or lying epitaph of the great. The former spring from and are
dictated by the impulses of nature and a desire to emulate, to hand
down the good name of their families as it descended to them. This
does them infinite honour. Few things provoke them so much as to
cast any reflection on their parentage. Knowing this, well indeed
would it be if all were to consider the full force of the saying, and
act so as to show the practical utility, with a view of conferring the
boon.

The paucity of crime in the mining districts of Cornwall has been
the subject of surprise, comment, and gratification too, to almost
every judge who has travelled the circuit. No better attestation of
this truth could be elicited than the number of convictions at the
County Assizes: they are found to be far less in proportion to the
number of inhabitants, and lowness of wages, than in any other
county in England. Outrage is almost unknown, the higher crimes
are extremely rare, and when perpetrated are generally by natives of
other counties resident in Cornwall. The people are thrifty, live on
hard, very hard, fare; are patient, quiet, and contented; value a
good name before everything else; are particularly proud of their
parentage, to a degree almost rivalling that of the Welsh, and refer to
King Arthur and Trelawney as demigods and patterns of virtue and
patriotism. The soul-stirring patriotic and favourite song of—

> And shall Trelawney die,
> These twenty thousand Cornishmen
> Will know the reason why,

is still sung by them, the chorus commanding the voices of the whole
company whenever introduced.

This little piece of pride (it is not vanity, it is just pride) should be cherished, it leads to many a virtuous action, lest the havage should be sullied, and their descendants have the reflection of that, perhaps, the only thing they have to bequeath on their deathbed should be tarnished. Many a spirit has departed to its long rest, at ease in the consciousness that, if it had nothing else to leave behind, it left to its posterity a good havage.

To so desirable a state of society the spread of religious instruction and Sunday schools have undoubtedly largely contributed; but we cannot help thinking that old Cornish pride, that innate and acknowledged hospitality and worth, have been in some degree engendered and fostered by the principle which enunciated the motto—"A clomen cat [13] is poor havage."

## Wheels within Wheels

THE APPLICATION OF THIS ADAGE, WE THINK, IS SO SELF-evident to mining undertakings in almost every clime, that lengthened comment would be almost superfluous, were it not that the adage is repudiated by every promoter of a mining speculation or adventure. Every one protests his to be free from anything like secret or unseen motive in his endeavour to favour the public; he asserts his to be purely a *bona fide*, disinterested, fair, and aboveboard scheme, yet if the glowing prospectus be carefully studied the little pinions on which the whole affair rests may be detected. It is generally found where the greatest endeavours have been made to hide the machinery the "wheels within wheels" are the most numerous and most extensive. How many gold schemes have had their "wheels within wheels" laid bare—aye, and Cornish, Welsh, indeed, all mines in all counties, have from time to time been exposed. We believe, however, fewer of the delusive schemes and self-inviting mines are to be found than ever was the case. Adventurers are too wary now-a-day to enter on speculations where all the speculation is on their part: the situations, the erection of machinery, sending in supplies, and free shares on the part of the promoters, we say we verily believe far less of that kind of practice is prevalent than formerly; the very magnitude of the practice worked its own cure, and a more wholesome state of things has been engendered and built up in their places. It is gratifying to find this to be the case for the very cause of mining itself, as it must tend to inspire confidence and extend its connections. We know many instances in which worthy and wealthy men have been induced to embark, but when they detected the "wheels" have thrown up their shares rather than be led into secrets that they knew not of. It must be admitted that mining has had its full share in such works; it must be also granted that it has had more than its share of stigma and odium. Now that a better system is adopted, it is to be hoped as liberal a policy will be pursued by capitalists, and that they will more readily embark in interests proved beyond all gainsay to be one of the best paying and most secure of all our national resources, employing the most industrious, patient, and deserving of our working population, whose endeavours not only maintain a vast number of their own class, but a large staff of able and intelligent superintendents; not only are these benefited, but shipping, manufacturers, and tradesmen, are all seriously affected by the prosperity or adversity of mining. The lords of the land, by their diminished dues, too suffer to some extent, but they and the farmer are almost to the same degree as the other classes enumerated. We, therefore, invite them to join, feeling assured they may now do so with comparative safety: the Augean stable has been swept; the harpies have been scared, or are known; the demand for the metals is good, and likely to be permanent; but before they rush into speculative undertakings to carefully examine them in all their bearings, to bear in mind our motto, and observe the "Wheels within wheels."

## GOOD MINES MAKE GOOD CAPTAINS

WHOEVER PERUSES THIS PAPER WILL ADMIT OF ITS GENERAL use in, and the truth of its application to, the mining districts at least. But its real application is universal; success, even by chance, is by the public too often deemed the proof of ability and the standard of merit. Prone is it to fall down and worship the golden calf—to judge of the tinsel instead of the sterling material. So often is it that a mine captain, if his mine, from no merit of his own, yields riches, is idolised; the crowd run after him; hosts of speculators seek his society, and court his favours; everything he touches turns to gold; everything he recommends is eagerly seized; all he says and does must be right, and all he condemns must be wrong. We are led to these remarks from having frequently—oh! how too frequently— witnessed the truth of them. How often have we seen the ignorant but lucky individual lauded to the skies, and laden with riches; whilst plodding industry and real talent have been, by untoward circumstances, doomed to tread the paths of distress, and eat the bitter bread of poverty—to suffer the language of unworthy reproach, and witness the finger of scorn? But so it is, so it has been, and so it will possibly always be; unpatronized merit must be neglected. Unless fortune, by some unexpected event, shower upon him riches, he has little chance. Many a hard-working miner has toiled all his life, and expended his days in dangerous employment, without meeting with any reward, save the recriminations of his employers; whilst his idle, drunken, next-door neighbour, perhaps by a discovery of which he had no idea, and could by no possibility have calculated on, receives the overwhelming compliments of those who know him not, and also those who know him. His failings now are not faults; his vices not realities, but only eccentricities, the marks of true genius, and he is treated accordingly. The plodding endeavours of the other are attributed to his extreme dullness, want of energy, and total unfitness for a business requiring high spirit and high mental powers.

If any proof of the truth of the maxim were required, the reader has only to look about him; if he does not recognise instances he is a rare individual indeed; he has only to apply to his neighbour, who will assuredly do so. Nothing can be so galling, nothing so dispiriting to a mine captain, or so disheartening to genius, as to be continually taunted with such comparisons by their adventurers, or by the misguided patronage of the public; it does infinite injury, but unfortunately that is unheeded. Man is not judged by his merits, but by his gold. His endeavours are never placed in the scale against so ponderous an article as the precious, over precious, metal; its glare obscures the sight of reason, and dazzles the eyes of the world; whilst the modest demeanour and shabby garb of ability only excites the disgust or ridicule of the inconsiderate. The russet coat frequently covers a much more honest heart, and ten times more ability, than the most showy paletot and polished exterior.

In mining this may almost be taken as a truism and an universal fact, by which guidance may be taken; for, be it remembered, Fortune is fickle. Indeed, it seldom happens that one captain discovers more than one good mine; though he may have been fortunate in one or two instances, it does not follow as a matter of course he is infallible, or his opinions invariably correct. No man can see a hole through a stone until it is made, nor can he see what is in the ground; his guide must, or should be, practice, analogy, and experience. The more he sees and knows of geology and the science of mining, the more deserving is his opinion of respect, though it be given by a non-successful practitioner; but we suppose it will ever be, as we have before observed, in practice—"A good mine makes a good captain."

## A Down Souse Man don't Cheffer

WE MET WITH THIS OLD AND QUAINT SAYING LAST WEEK, on witnessing a bargain between a buyer of tinstone and a miner who had raised it on a tribute venture. The tinner, who well knew his business, asked a fair price for his commodity, whilst the old, shrewd tin dresser, who also knew the true value, endeavoured to purchase it as cheaply as possible. The miner refused to sell at anything less than he at first named; and, after a good deal of badinage on both sides, the miner turned round on his heel, and, uttering the exclamation at the head of our paper, with a look of the most ineffable contempt, that rendered it doubly sarcastic and significant, told the dresser he would not now sell it to him at any price; "For," said he, "had you the chance you would take me in. I likes a 'down souse man.'"

How few "down souse men" are to be met now-a-day, is the moral to be derived from the adage; and how great contempt are those looked upon by their fellows who are not? How much would business transactions be simplified and confidence established by adopting the principle enunciated in the homespun, antiquated maxim of our miner?

There is much room for reflection on a subject like this, and we should all do well to study it in its simplicity. "Down souse" to any but a Cornish ear sounds oddly, but the very pronunciation of the words are indicative of their meaning, many of the old Cornish words having the same peculiarity of expression,—by the sound you may judge of their real significance; this, we think, is a favourable example, and we shall from time to time make such quotations from the nearly obsolete Cornish language as will show the idiomatic nature of its construction, as in former use. They are now only to be met with in the old sayings, and even these in remote localities. Still they should not be lost; and we wonder no resident Cornishman has not thought it worth his while to redeem them from oblivion, many being rich in genuine humour, *double entendre*, and sarcasm. That forming the subject of our paper is one of sterling, downright, straightforward honesty and bluntness, such as we can easily suppose would be uttered by an unsophisticated son of labour, who perceived he was being trifled with by a practised man of the world, and who, in his conscious integrity, would exclaim—"A down souse[14] man don't cheffer.[15]"

## WHAT WON'T MAKE TIN WILL TAKE TIN

HERE IS ANOTHER OF THESE QUAINT SAYINGS THAT HAVE A double meaning, each of which has an equally forcible application. To the miner it is well known that in dressing tin ores it is necessary the stone should be so selected as, if possible, to allow no heavy waste, or other objectionable matter, known to experience only, to be mixed with it, as it will inevitably cause loss, not only of time and trouble, but will actually so contaminate the ore that separation is next to impossible; hence the peculiar origin of the sentence. But we see it may be applied with equal advantage to everyday life. If time and ability be not profitably employed, it may be depended on the reverse will be the fact.

> "Satan finds some mischief still,
> For idle hands to do."

The application, it will thus be seen, is clothed under a homely guise, much better adapted to the understandings of the parties to whom they were addressed than more elaborate and polished lessons. These maxims also prove that the Cornish people, at the remote period of their introduction, entertained fine views of the utility of proverbs, for we see, in all instances, though they contain great wit, and frequently severe criticism, they invariably have a moral at the bottom. In this instance, the moral is distinctly seen by the Cornish miner, as well as its derivation from his daily occupation. It is a goodly motto; and it will be well for society generally, as for the parties to whom it is addressed, to remember that "What won't make tin will take tin."

### 'HOLD THY JAW, DO,' SAYS JOHN TREGONING

THIS RATHER VULGAR EXPRESSION IS BUT A REFLEX OF THE idea expressed some thousand years since by the poet Terence, in his proverb of *Ne sundas narrat fabulam*, "He knows not to what a deaf ass he talks." We should not have quoted the Latin apothegm, but to show the same animus is expressed in all ages in pretty nearly the same language. The John Tregoning alluded to was one of the old school of Cornish mine captains who, by dint of attention, raised themselves to fame, wealth, and distinction amongst not only their fellows, but are looked up to even by their superiors, yet still retain their original bluntness and coarseness of demeanour and language. This is too frequently cherished by them as a distinguishing feature, and is often regarded by strangers as a certain sign of originality and independence. Fortunately this feeling is on the decline on both sides, it being found that the gentleman of education, fine feeling, and polished language is not incompatible with the situation of mine manager or captain.

As we have said, the authority by whom the adage was quoted so frequently as to become a sort of addendum to the original was one of the class of captains nearly extinct. The saying is now confined to the lower class; still it is prevalent, and the ear is so frequently saluted with the sound that we sometimes are instinctively tempted to exclaim, when we hear persons who pretend to understand mining, for such there be (particularly in the metropolis), " 'Hold thy jaw, do,'[16] said John Tregoning." On such occasions was it the old 'cute miner was accustomed in polite society to utter the curt sentence.

The chatter of a magpie, the jargon of a parrot, or the rattle of a scolding woman's tongue, can scarcely be more annoying to an old experienced miner than the language and dictation to him of how to work a mine, frequently vouchsafed by the committee through their Chairman—the members of the said committee probably never having seen or known what mining is or should be, guided, as is frequently the case, by the mystification of a multitude of opinions and reports, or by a wretched and ruinous parsimony. When dictate to in such a manner, is it any wonder that the man of experience is ready to cry out in his vexation, "Hold thy jaw, do?"

It is much to be regretted that instances so much resembling the case above described do really so frequently recur; it is one of the consequences of the defective state of mine management. Too often is patronage, not ability, the stepping-stone to situations for which the employed are both physically and practically wholly incompetent; the natural consequences result; the affair gets into difficulty and disrepute; the mine becomes (though intrinsically valuable) a ruinous concern, the shareholders disgusted, and the agents dismissed. When the proprietors, deeming that in a multitude of counsellors there must be much wisdom, depute the entire management to a body constituted by them, who sit in conclave, and issue their orders from the "office," that the board will sanction nothing but so and so, that the expenses must be limited to so much per month. Oh! ye sages, could you but hear as we do the to you silent, but to us audible and significant, "Hold thy jaw, do," you would be more cautious on behalf of your own as well as your brother adventurers' pockets than you are of how you issue your edicts to direct experience and instruct ability.

To apply our rough subject to advantage from the above true statements, let us advise a contrary mode of procedure. Advance merit and ability to their legitimate positions, and let patronage of "Dowbs" cease in mining, at all events; it is not an employment for pedantic fops or Cockney amateurs—let mining be to miners. By doing so our phrase would grow into disuetude, and ultimately sink into oblivion, save that this sketch may rescue it as being amongst the list of Cornish proverbs, many of which are obsolete. Would this were so, as well as the practices to which it refers, and that we hear no more from captains' mouths in their vexation, " 'Hold thy jaw, do,' says John Tregoning."

## The Mayor of Market Jew

LIKE MOST MINING PARTS OF GREAT BRITAIN, THE VARIOUS towns as well as individuals in Cornwall are known by some sobriquet or nickname. Whence this ancient custom arose we know not, but we find it prevalent in Lancashire, Wales, and the North of England, and in all probability this practice has been the origin of the great variety of surnames. Be this as it may, the cognomen of many families frequently display a singularity, quaintness, or witticism, that bears strong testimony in favour of such an hypothesis. "The Mayor of Market Jew" is, as many be inferred from the above remarks, a myth, a sort of badinage, used by the inhabitants of the neighbouring villages to ridicule the population of that ancient market town; but the addendum to the apothegm is so applicable for general use that it has become a complete axiom throughout the county at large. We, therefore, insert it for the benefit and amusement of the public generally, and let them enquire whether or not among their own sphere they can discover a "Mayor of Market Jew," or *a man who sits in his own light*, for he is the veritable "Mayor of Market Jew."

It may not be uninteresting here to give a brief history of this village, which was undoubtedly once of far greater consequence than at present, though it has never been endowed with corporate honours. It is said to have been one of the principal entrepots of the ancient tin trade of Cornwall—that the Jews and Phoenecians frequented it for the purchase of that mineral, and the sale of their various commodities. The famous St. Michael's Mount and subjacent rocks are by many supposed to have been the Cassiterides, or islands of tin, alluded to by Strabo, Pliny, and other authors. Though the Islands of Scilly claim the distinction, certainly the position of Market Jew, or Jew Market, as well as its name, implies a strong argumentative presumption in its favour. The place is in the immediate locality of extensive stream works in the parishes of Sancreed, Madron, Morvah, Gulval, Ludgvan, St. Erth, Germoe, &c.

The town at present is known as Marazion. Its principal support is still from the adjacent mines, the chief one of which is the celebrated Tolvadden, or the modern Cornish mining wonder. Near it is the Great Wheal Fortune, once so exceedingly rich as to have returned more than a million of money. The Marazion mines, the old Neptune, and many others, might be enumerated, but they cause us too great a digression.

Our Maxim, in brief and homely phrase, expresses a most important consideration—a man sitting in his own light; or, in more erudite language, by his own waywardness, self-will, and folly, defeating his own ends, disappointing his own expectations, and overreaching himself. Most of these quaint adages are admirably adapted to the cant idiosyncracy of the Cornish people, who avoid anything like detail if they can convey the same ideas in brief sentences, particularly when a smack of sarcasm is engrafted. We believe the maxim is confined to the county of Cornwall, where it is well known; but, as we have before hinted, it by no means expresses the limits of the sphere of action of "Mayors of Market Jew."

These phrases are exceedingly useful amongst such a population as that by whom they are used. Many are becoming obsolete; this is one of the most popular, most extensively used, and, therefore, best preserved. And now let us ask the reader, has he in his experience never met with a character of this description? If not, verily he is a singular man. We confess to have done so frequently, and yield to the soft impeachment of having more than once been in the predicament; but if our readers have not yet seen or been such, let them read, mark, and recollect the old Cornish maxim—"The Mayor of Market Jew."

## TOO MANY COOKS SPOIL THE BROTH

BY MANY HYPERCRITICS OUR CLAIM TO THIS MAXIM AS A Cornish mining idiom will be looked upon as trespassing out of our legitimate limits, but to the Cornish miner its adaptation is peculiarly fitting and appropriate. In none of the whole round of trades and professions is it more applicable, nor in them are there to be found a greater number of professors, whose *quasi* abilities are more loudly trumpeted forth to the world, or any in which private advice, by the same caterers for the public appetites of novelty and scandal, are more assiduously and greedily sought. Let us try our hand at actually practised and every-day facts, and see if we cannot fully establish our claim.

Instances have so frequently come under our notice that we doubt not all who are engaged in mining affairs have witnessed the baneful effects of the variety of opinion—some actuated by the desire of being thought more able because differing, some from a mere innate spirit of contradiction, but by far the greatest number from the hope of personal benefit. There is yet another great source of the success of this species of interlopers. Sanguine adventurers too frequently rush into speculations from that unfortunate but instinctive motive which makes man, like all gregarious animals, follow their leader, and where the rush takes place they, of course, rush too, even though they be warned of the consequences. Ruin, as a matter of course, follows. In their distress they, as a forlorn hope, call in the advice of one of the "cooking" fraternity, who, knowing his business (like a peripatetic quack), flatters the consulting party, by giving him his own way, and fanning the growing flame already heating his brain, depressing his energy and spirits. So prone are men generally to listen to any opinions that are in harmony with their own, that they dub those who coincide as very clever fellows: they lavish patronage and favours with a liberal hand on the fawning hypocrite; whilst the plain-spoken, honest, upright man, though doing his best for his employers, is disbelieved, and treated at first with neglect, followed by indignity and contempt, as the insidiously distilled poison takes effect on the overweening, unconscious victim.

By these means the best constituted mining companies have been ruined. Let a disappointed shareholder but once consult one of these gentry, and he will cook a mess out of a mine as surely as a lawyer will make a Chancery suit out of a disputed will, or a doctor make a bill out of a rich hypocondriac. To the pedantic pedagogue a wealthy client is an invaluable windfall. All the selections he has made, and all the investments he may anticipate making, have been conceived in error, and carried out with imbecillity.

In the working department of a mine the mischief is still worse; for here the same principles act with tenfold violence, and the slightest fault, real or supposed, on the part of the manager or agent, is magnified into the most tremendous villainy by the poisoned mind. We could instance one case in which a sawyer on a mine was apprehended by the cooking manager for carrying home a few chips in his dinner basket, who, by the aid of his cooking apparatus, had represented in his report to the shareholders at Leeds that a systematic robbery of timber was being carried on, and that he had actually detected sundry pieces of valuable wood being conveyed from the mine. Clever fellow this, of course. The captains and agents were blamed, and by the poisoned adventurers were boldly charged with either winking at or encouraging such nefarious doings; whereas, a strict investigation which was made exposed the truth, and the cook was unmasked: but had the "leprous distilment" fallen on the ear of a less energetic man, the consequences might have been disastrous; the adventurer might have silently disposed of his interest at a ruinous sacrifice, too happy to have escaped from a nest of thieves.

We might multiply and vary the proposition and adaptation of our motto *ad infinitum;* the doing so is so simple, self-evident, and easy, that we leave the task, and beg our readers to apply the moral of the proverb; they may rely on its truth. Mystification in mining is sufficiently prevalent without the aid of cooks, particularly "too many." How much would have been spared to East Russell had cooks been discarded? How many pangs and heart-burnings would have been saved—how many angry epistles have been left unpenned, had this moral been acted on? We pen the Maxim on this occasion, as we know of a property really valuable that is on the eve of destruction from this evil, cautioning them that one good cook is sufficient in a kitchen, and one good tailor enough for one suit—that the axiom is good in their case peculiarly, that it has been proved so in general practice, and that they should take care how they proceed; to learn from experience, and take heed in time. We hope they will, as is intended they should, profit by our Maxim, which is as much Cornish as English, that "Too many cooks spoil the broth."

## ONE AND ALL

THIS MOTTO, THE WATCH WORD AND BATTLE CRY OF THE Cornish, is of great antiquity, as is proved by inscriptions of very remote dates. No doubt it originated in that peculiar characteristic of the Cornish people still existing in a great degree, notwithstanding the extended intercourse and familiarity with strangers consequent on modern improvements in society. Formerly the opportunity of visiting the metropolis was considered by a Cornishman as an event in his life; on his return he became quite a "lion" of the district, and an oracle of wisdom; even at this day, in the rural districts, at a cobbler's stall may be seen the important announcement—"from London." This complete isolation of the inhabitants rendered intermarriages of families the rule, hence the familiar term of cousin, uncle, and aunt, so continually heard in the county, where the people are caricatured by strangers as cousin Jackey, uncle Jan, and aunt Jenny; but the true feeling, idiom, and meaning of our motto will afford to forgive all these little witticisms and familiarities of "foreigners", as "up the country people" are sometimes called. So applicable, indeed, is the county motto to the habits and dispositions of the Cornish people that we doubt if a phrase more suitable could by possibility be discovered. Curious coincidences of mottoes suiting family characteristics, in a similar manner, are sometimes to be met with, as in the case of the Napiers, where "Ready, aye Ready" adorns their banner. The tantamount meaning of the Cornish maxim in the catalogue of English apothegms would be "Union is

Strength." Probably the Master of Heraldry, who adopted it in his day, had this sentence in view when arranging the arms of the Duchy and county where the fifteen bezants are placed in a triangular form as indicative of union and strength. Be this as it may, the practical illustration of the maxim is necessary to the development of all the natural resources of the county, in which few opportunities are offered for success by individual enterprise, whilst in no part of the world is greater scope afforded for undertakings in which combination with a determined spirit of definite action are necessary.

This remark holds good in all her principal natural resources, whether mining, fishing, or her immense nautical marine be considered; combination and unity of action must be the essential to the execution, on a proper scale, of either of these. Nature seems to adapt the natives of every clime and every locality to the necessities of their fatherland. The Cornish are remarkable for their sanguine temperament, their indomitable perseverance, their ardent hope in adventure, and their desire for discovery and novelty; hence their wide distribution all over the world, in the most remote corners of which they are to be found amongst the pioneers; and to this very cause has science to boast of so many brilliant ornaments who claim Cornwall as their birthplace.

But we have more immediately to do with our subject as relating to mining, and here we beg again to remark that nothing could be so suitable to our purpose, as unity of action is the mainspring of success in all undertakings of this description; more mining companies have to date their ruin from want of attention to this precept than from all other causes put together, numerous and various as they are. This motto should be so much a household word that every account-house should have it painted over the door, and every board-room should have it displayed in large letters, to which a Chairman might point when he is surrounded by quarrelling and disputing shareholders.

Amongst the mining population the sentence is still a rallying word; "One and All" is the sound by which they may be at any time led, not driven. To the latter purpose we hope it will never have to be applied, but that in every day practice, whether in combination to work and carry out a great undertaking of mining, fishing, shipping, or other enterprise, for national or local benefit, the language of the Cornishman will still and for ever be "One and All."

OUT OF THE WORLD, AND DOWN TO ST. IVES
THIS WELL-KNOWN SAYING AFFORDS A MOST EXCELLENT
practical demonstration of the folly of prejudice, particularly in
mining matters. The ancient borough to which allusion is made was
formerly out of the prescribed limits of mining influences. The few
mines in the locality were treated as being, and virtually were,
wrought for mere boroughmongering and electioneering purposes;
the whole district was held to be literally out of the mining world
(the "world" of the Cornish), and to be only worthy of consideration as a rotten pocket property, whose voters were venal enough
to be purchased by any adventurer bidding the highest; and as a
fishing village, of no worth but for pilchards and train oil. How
often, however, does the rejected stone become the head of the
corner. We hear now no more of "Out of the World, and down to
St. Ives;" but, if you want tin go to St. Ives, the mines are among
the most prominent in this branch of Cornwall's produce, and the
district one of the most popular and attractive to speculators. Well-
conducted and carefully-managed properties have led to this great
and interesting change. As long as they were under the influences of
venal strangers the vast worth of these mines was hidden; but when
liberated from the thraldom of party purposes, and the curse of
prejudice, and were consigned to the speculative spirit of real miners,
their industry, perseverance, and science soon established them as
among the most excellent properties the county of Cornwall can
boast, enriching her landowners by revenues a prince might envy,
creating a happy, contented working population, and scattering all
those blessings that are sure to follow in the train of successful mining
adventure, in addition to the intrinsic value added to the common
weal of the kingdom by the immense profits to the proprietary, and
the impulse to commerce thereby created.

Such are the facts of the case; the inhabitants of St. Ives know and
feel that the value of their mining property is of far greater consequence than the mere patronage of a few talkative, pretending
patriots, who only visited and flattered them on the eve of an
election, and too frequently left the players themselves to pay the
pipers. They know, and feel too, that the bar of extraction is taken
off, and that this once almost (in Cornwall) universal maxim is fast
becoming obsolete.

In mining it teaches, or should teach, us not to condemn from
hearsay, prejudice or tradition; that under the benign influences of
capital, science, and perseverance, many spots that have been and are

lingering under the most unfortunate circumstances may, like the one referred to, become, if not the greatest, an integral portion of the community of which it is a member. Although the phrase has become obsolete, the memory of it should be cherished; to the adventurer and young miner it may remain in full force, and in the hour of need they should remember "Out of the world, and down to St. Ives."

[1] Berrin—a funeral. The churches in Cornwall being so far apart, and no hearses obtainable by the humbler classes, these assemblages of miners are absolutely necessary. Their funerals, therefore, if possible, take place on a Sunday. It should be observed that these men hold a peculiar sanctity for the dead—*De mortuis nil nisi bonum* is literally their feeling.

[2] Scale—a piece of loose detached ground falling. These accidents frequently happen without the slightest warning.

[3] Lashed—knocked or struck down.

[4] Whish't—melancholy, wretched. A curious term, but very expressive when pronounced with a melancholy accent. Thus, a miserable wet day in Cornwall is termed "a whish't day."

[5] Comrade—the usual acceptation of this word does not quite come up to the Cornish, who mean something more than bare companion—it means bosom friend.

[6] Bearers—The coffin is always carried underhand; and in such long distances, frequent relays are absolutely necessary.

[7] On the top of the tower at St. Michael's Mount are the remains of an old stone lantern, in which a light was formerly kept as a beacon for mariners. It is now disused, and the upper part gone, thus forming a rude kind of chair, in which the sitter dangles his feet some hundred or two feet high. The slightest slip would cause a fall, and certain destruction. Strange to say, no fatal accident has been recorded, though the place cannot be viewed without a shudder. The object is a silly tradition, that whoever goes first into the chair remains the governor after marriage. This foolish would-be witticism is the source of great glee, as may be supposed, on such occasions.

[8] Smuggled brandy and hollands.

[9] It is said the devil never comes into Cornwall, for fear of being served as everything else is in that County.

[10] Trifling luxuries.

[11] Carglaze is said to mean a view from a hill, as Pentireglaze means a view from an headland.

[12] Deep, in Cornish parlance, means cunning as well as clever.

13 About 20 years since in the cottage of every miner might have been seen one of those cats made of plaster of Paris vended by vagrant image sellers. They seemed almost to be their *penates*. At present they are superseded by tawdry Staffordshire ware representations of celebrities, amongst whom the Queen and Prince Albert, with the old Duke, seem to be the greatest favourites; after them the Duke of Cornwall takes pre-eminence. Of gentle shepherds and their mates, in every possible and impossible variety and costume, there are incredible quantities. These people appear to have a taste for such decorations: what a pity but opportunity were afforded them of cultivating it by better examples in the ceramic art.

14 "Down souse," determined, plain-spoken.

15 "Cheffer," dispute, haggle.

16 Hold your tongue, do; or, you don't know what you are saying.

# INDEX

## INDEX OF MINES

THE MINES AND MINERAL RAILWAYS OF
EAST CORNWALL AND WEST DEVON    D. B. BARTON
A historical guide to the mines of east Cornwall and west Devon, outlining
their position and history, together with that of the mineral railways that served
them. [Second edition, 1972]

LEVANT: THE MINE BENEATH THE SEA    CYRIL NOALL
Levant Mine, on the wild Atlantic cliffs by St. Just, occupies a unique niche in the
annals of Cornish mining. Its history, told here in full for the first time, made it an
obvious choice for the initial volume in the new *Monographs on Mining History*
series.

THE CORNISH BEAM ENGINE    D. B. BARTON
A survey of its history and development in the mines of Cornwall and Devon
from before 1800 to the present day. Chapters are included on engines used not
only for pumping but also for winding and stamping ore, for working man-
engines, etc. [Second edition 1969]

A HISTORY OF THE CORNISH CHINA-CLAY INDUSTRY
R. M. BARTON
The definitive history of the china-clay and china-stone industry of Cornwall—
the most important and individual extractive industries in Great Britain today.

A HISTORY OF TIN MINING AND SMELTING
IN CORNWALL    D. B. BARTON
An account of tin mining and smelting in Cornwall—both hitherto unchronicled
as industries—from 1800 to the present day. A book which puts the past into
perspective and the present into focus.

ON THE STEAM ENGINES IN CORNWALL    THOMAS LEAN
Lean's *Engine Reporter*, from which this book was compiled in 1839, was a nineteenth century publication unique in the annals of steam engineering history. This
reprint of a rare and classic work will be welcomed by all interested in this subject
and in the Cornish beam engine in particular.

HISTORIC CORNISH MINING SCENES UNDERGROUND
Edited by D. B. BARTON
Photographs taken underground in metal mines are rarities, particularly those
taken before the turn of the century. The historic value of the plates in this book
thus need little stressing, for they form a unique collection.

D.  BRADFORD                    BARTON    LTD
TRETHELLAN HOUSE                TRURO CORNWALL